Springer Theses

Recognizing Outstanding Ph.D. Research

Aims and Scope

The series "Springer Theses" brings together a selection of the very best Ph.D. theses from around the world and across the physical sciences. Nominated and endorsed by two recognized specialists, each published volume has been selected for its scientific excellence and the high impact of its contents for the pertinent field of research. For greater accessibility to non-specialists, the published versions include an extended introduction, as well as a foreword by the student's supervisor explaining the special relevance of the work for the field. As a whole, the series will provide a valuable resource both for newcomers to the research fields described, and for other scientists seeking detailed background information on special questions. Finally, it provides an accredited documentation of the valuable contributions made by today's younger generation of scientists.

Theses are accepted into the series by invited nomination only and must fulfill all of the following criteria

- They must be written in good English.
- The topic should fall within the confines of Chemistry, Physics, Earth Sciences, Engineering and related interdisciplinary fields such as Materials, Nanoscience, Chemical Engineering, Complex Systems and Biophysics.
- The work reported in the thesis must represent a significant scientific advance.
- If the thesis includes previously published material, permission to reproduce this must be gained from the respective copyright holder.
- They must have been examined and passed during the 12 months prior to nomination.
- Each thesis should include a foreword by the supervisor outlining the significance of its content.
- The theses should have a clearly defined structure including an introduction accessible to scientists not expert in that particular field.

More information about this series at http://www.springer.com/series/8790

Jiandong Sun

Field-effect Self-mixing Terahertz Detectors

Doctoral Thesis accepted by
Chinese Academy of Sciences, China

Author
Dr. Jiandong Sun
Key Laboratory of Nanodevices and Applications
Suzhou Institute of Nano-tech and Nano-bionics
Chinese Academy of Sciences
Suzhou
China

Supervisor
Prof. Hua Qin
Key Laboratory of Nanodevices and Applications
Suzhou Institute of Nano-tech and Nano-bionics
Chinese Academy of Sciences
Suzhou
China

ISSN 2190-5053 ISSN 2190-5061 (electronic)
Springer Theses
ISBN 978-3-662-48679-5 ISBN 978-3-662-48681-8 (eBook)
DOI 10.1007/978-3-662-48681-8

Library of Congress Control Number: 2015959929

Printed on acid-free paper

This Springer imprint is published by SpringerNature
The registered company is Springer-Verlag GmbH Berlin Heidelberg

Parts of this thesis have been published in the following journal articles:

1. **Sun, J.D.**, Sun, Y.F., Wu, D.M., Cai, Y., Qin, H., Zhang, B.S.: High-responsivity, low-noise, room-temperature, self-mixing terahertz detector realized using floating antennas on a GaN-based field-effect transistor. Appl. Phys. Lett. **100**, 013506 (2012)
2. **Sun, J.D.**, Qin, H., Lewis, R.A., Sun, Y.F., Zhang, X.Y., Cai, Y., Wu, D.M., Zhang, B.S.: Probing and modelling the localized self-mixing in a GaN/AlGaN field-effect terahertz detector. Appl. Phys. Lett. **100**, 173513 (2012)
3. **Sun, J.D.**, Qin, H., Lewis, R.A., Yang, X.X., Sun, Y.F., Zhang, Z.P., Li, X.X., Zhang, X.Y., Cai, Y., Wu, D.M., Zhang, B.S.: The effect of symmetry on resonant and nonresonant photoresponses in a field-effect terahertz detector. Appl. Phys. Lett. **106**, 031119 (2015)
4. Lü, L., **Sun, J.D.**, Lewis, R.A., Sun, Y.F., Wu, D.M., Cai, Y., Qin, H.: Mapping an on-chip terahertz antenna by a scanning near-field probe and a fixed field-effect transistor. Chin. Phys. B. **24**(2), 028504 (2015)
5. **Sun, J.D.**, Sun, Y.F., Zhou, Y., Zhang, Z.P., Lin, W.K., C.H., Zeng, Wu, D.M., Zhang, B.S., Qin, H., Li, L.L., Xu, W.: Enhancement of terahertz coupling efficiency by improved antenna design in GaN/AlGaN HEMT detectors. AIP Conf. Proc. **1399**, 893 (2011)
6. Sun, Y.F., **Sun, J.D.**, Zhou, Y., Tan, R.B., Zeng, C.H., Xue, W., Qin, H., Zhang, B.S., Wu, D.M.: Room temperature GaN/AlGaN self-mixing terahertz detector enhanced by resonant antennas. Appl. Phys. Lett. **98**, 252103 (2011)

Dedicated to my parents, my wife, and my daughter

Supervisor's Foreword

Searching of efficient solid-state terahertz devices including terahertz emitters, detectors, and modulators has been one of the most challenging tasks to bring various terahertz applications into reality. Terahertz plasmonic devices based on low-dimensional semiconductor structures allow for active tuning of the electronic states, terahertz electromagnetic wave and the coupling in between. Being a collective mode rather than a single-particle state, plasmon in semiconductors may provide solutions to high-sensitivity terahertz detection and high-efficiency terahertz emission.

In the August of 2008, two Ph.D. candidates Jiandong Sun and Yu Zhou under my supervision initiated their research on terahertz devices by using plasma waves in AlGaN/GaN two-dimensional electron gas (2DEG). The initial idea was to follow up with Dyakonov–Shur's proposals on resonant plasmon excitation in 2DEG either by an incident terahertz electromagnetic wave or by a driving electrical current, two approaches to realize terahertz detectors and terahertz emitters, respectively. At the first attempt Dr. Yu Zhou succeeded in observing terahertz photo current in a conventional AlGaN/GaN high-electron-mobility transistor (HEMT) at 0.9 THz. Since then, Dr. Jiandong Sun has started his systematic studies on terahertz detection by using antenna coupled field-effect transistors and Dr. Yu Zhou has started to search for new methods for efficient terahertz emission from 2DEG plasmons.

The observed photocurrent came from self-mixing of terahertz electric field or non-resonant excitation of plasma wave in the field-effect electron channel. Dr. Sun et al. realized that the self-mixing effect allows for high-sensitivity terahertz detection at room temperature. His thesis focuses on the modeling and verification of the self-mixing mechanism. By performing device modeling/simulation, detector design, fabrication and characterization, systematic and original results were obtained to gain deep understanding on the device physics. The thesis is comprehensive in the aspect of the plasmonic device physics and the characterization techniques. The findings are strongly supported by the techniques he developed, e.g., to probe the localized photo current in the micrometer-sized electron channel and to image the terahertz near-filed response of the antennas. The thesis is organized in the following seven chapters.

Chapter 1: Terahertz technologies in general, the terahertz sources and detectors are briefly introduced. The outline of this thesis is sketched.

Chapter 2: Based on a short review of the previous theories on terahertz plasmon detection, a quasi-static self-mixing detector model is developed by taking into account the spatial distributions of both the terahertz electric field and the electron density manipulated by antennas and the gate.

Chapter 3: Experiments on the design, fabrication, and characterization of self-mixing detectors based on AlGaN/GaN 2DEG are presented. Optimization of the detectors guided by the model is also introduced.

Chapter 4: Observation of resonant plasmon detection in a detector with symmetric antennas and nano-gates at 77 K. The symmetric design allows for the distinguishing of resonant plasmon detection from the non-resonant self-mixing detection.

Chapter 5: A scanning probe and the field-effect electron channel are used to probe the near-field properties of the terahertz antennas. The asymmetric distribution of terahertz field is confirmed.

Chapter 6: Self-mixing terahertz detectors are tested for terahertz imaging and Fourier transform spectroscopy.

Chapter 7: Conclusions and outlook are given.

This thesis covers a broad scope of self-mixing terahertz detection including the physical detection model, device simulation, fabrication and characterization. From the point of view of device physics, the most valuable finding of this thesis to the community is the detector model which although is of quasi-static describes clearly the effect of symmetries in the spatial distributions of the terahertz electric field and the electron density. He verified the model by manipulating the spatial distributions of the terahertz field and the electron channel using asymmetric/symmetric antennas and the in-plane/perpendicular DC electric fields, respectively. Furthermore, he successfully demonstrated the optimization of the sensitivity by improving the integrated terahertz antenna and the gate. He designed an antenna with three dipole antenna blocks and realized strongly localized and spatially asymmetric terahertz field in the gated electron channel. The effectiveness of this design has been proven in self-mixing detectors based on AlGaN/GaN 2DEG and graphene (not included in this thesis). Dr. Sun also pointed out that non-resonant self-mixing is more pronounced than the resonant plasmon detection at room temperature by distinguishing these two different effects in a symmetric device.

This thesis also raises a few questions to be answered in the future. Among them there are a few closely related to the ultimate limit in sensitivity of self-mixing detection.

1. The detector model is of quasi-static without considering the dynamics of electron transport driven by the terahertz electric field. What is the fundamental sensitivity limit for self-mixing detection?

2. Can a self-mixing detector sense incoherent terahertz radiation from a black-body? If yes, what would the sensitivity be?
3. What would be the right way to achieve resonant plasmon detection at room temperature and an even higher sensitivity than the non-resonant self-mixing detection?

To answer these questions, new antenna design is required and also the model should be more rigorous by taking into account the dynamic electron response to the terahertz electromagnetic field and the impedance effect of the antenna coupled to the 2DEG. This thesis could provide a strong basis for further investigations.

Suzhou
July 2015

Prof. Hua Qin

Acknowledgments

I wish to acknowledge everyone who supported me to make this thesis complete and solid. First and foremost, I would like to express my deep gratitude to my supervisor, Prof. Hua Qin. It is his open-minded, enthusiastic, and rigorous attitude in research that encouraged me to challenge the difficult scientific and technical problems. It would not be possible for me to complete the thesis without his boundless support, keen insight, and guidance. I am very glad to meet a good teacher like him at the most important stage in my career. I also greatly appreciate many supports from Prof. Baoshun Zhang, Prof. Dongmin Wu, and Prof. Yong Cai. They have generously provided me insightful comments and with understanding and precious knowledge of semiconductors, physics, and fabrication techniques. I also would like to take this opportunity to extend my thanks to Prof. Roger Lewis who provided me many insightful comments on my experiment results and corrected my manuscripts word by word.

I am also very grateful to Dr. Yunfei Sun with whom I collaborated during the whole period of this thesis work on the simulation, design, fabrication, measurement, and analysis of the terahertz detectors. There are also many colleagues to name for appreciation. They are Dr. Xiaoyu Zhang for his help in microwave measurements, Dr. Shitao Lou for introducing me to the terahertz near-field measurement, Xinxing Li for supporting me with the low-temperature measurement, Zhipeng Zhang for helping me on the device fabrication, and Dr. Renbing Tan for discussion on electron transport in 2DEG. I am grateful for the friendships I share with Dr. Hongxin Liu, Dr. Li Lv, Dr. Yongdan Huang, Peng Chen, Daigui Liu, and all other group members in the Terahertz Laboratory.

It would not have been possible for me to undertake this doctoral thesis without the support from my family. I would like to give my special gratitude to my parents for their unconditional and endless support, inspiration, and love. They support me both spiritually and financially to ensure my achievements in Ph.D. studies. I thank my wife Lingling Pan for her love, prayers, and immense support throughout my Ph.D. journey. Without her encouragement and understanding, it would be impossible for me to finish this work. I am so blessed to have her and my precious daughter, Rui Sun, who is our pride and joy. I would like to express my heartfelt thanks to my sisters, Jianying Sun and Yaoying Sun, who supported not only me but also my whole family during this challenging period of time.

Suzhou
July 2015

Jiandong Sun

Contents

Acronyms

2DEG	Two-dimensional electron gas
AL	Antenna length
ALD	Atomic layer deposition
AW	Antenna width
BFOM	Baliga quality factor
BS	Beam splitter
BWOs	Backward wave oscillators
CD	Control detector
CW	Continuous wave
DC	Direct current
DQW	Double-quantum-well
EBE	Electron beam evaporation
EBL	Electron beam lithography
FDTD	Finite-difference time-domain
FELs	Free electron lasers
FETs	Field-effect transistors
FTS	Fourier transform spectrometers
FWHM	Full width at half maximum
GHz	Gigahertz
GCA	Gradual channel approximation
HEB	Hot electron bolometer
HEMT	High-electron-mobility transistor
IBE	Ion beam etching
ICP	Inductively coupled plasma etching
IMPATT	Impact avalanche and transit-time diode
IR	Infrared
JMF	Johnson quality factor
LNA	Low-noise amplifier
LR	Linear regime
MODFETs	Modulation-doped field-effect transistors
MOSFETs	Metal–oxide–semiconductor FETs

NDC	Negative differential conductance
NEP	Noise equivalent power
OAPs	Off-axis parabolic mirrors
OD	Original detector
QCLs	Quantum cascade lasers
RF	Radio frequency
RIE	Reactive ion etching
RTA	Rapid thermal annealing
RTD	Resonant tunneling diode
SBD	Schottky barrier diode
SR	Saturation regime
STJ	Superconducting tunnel junction
THz	Terahertz
THz-QWP	Terahertz quantum-well photodetectors
THz-TDS	Terahertz time-domain spectroscopy
TPX	Polymethylpentene
TR	Transition regime
UVL	UV lithography

Chapter 1
Introduction

Abstract In this chapter, the basic characteristics of terahertz electromagnetic wave and terahertz applications are briefly introduced. Challenges in general, to develop solid-state terahertz devices including emitters and detectors are discussed by comparing the existing principles and actual devices. The focus of this thesis is then refined to the understanding of the self-mixing mechanism and the development of practical detectors based on field-effect transistors or those alike. The outline of this thesis is given at the end of this chapter.

1.1 Brief Introduction to Terahertz Electromagnetic Wave

Terahertz ($1\,\text{THz} = 10^{12}\,\text{Hz} \sim 1\,\text{ps} \sim 300\,\mu\text{m} \sim 33\,\text{cm}^{-1} \sim 4.1\,\text{meV} \sim 47.6\,\text{K}$) radiation falls in a band of electromagnetic wave from 0.3 to 3 THz, which is in between the infrared and the millimeter wave band [1–3]. Unlike other frequency bands such as radio waves, microwave, infrared ray, visible light, ultraviolet ray, and gamma ray in which many applications and technologies can be found in daily life as shown in Fig. 1.1, the realm of terahertz science and technology is less explored and developed. Nowadays, a few terahertz applications are maturing and playing more and more important roles. Among these are the detection of terahertz wave as an important tool for discovering the universe, security screening for threats, terahertz time-domain spectroscopy (THz-TDS) in a variety of tools for material science, biomedicine and sensing, etc. Many more exciting applications are expected to be realized. However, the advance of terahertz technology vastly relies on the development of terahertz devices including emitters, detectors, and modulators. Compact solid-state terahertz devices for room temperature operation are mostly preferred. Tremendous difficulties have been encountered in the development of solid-state terahertz devices for efficient generation, manipulation, and detection of terahertz electromagnetic wave at room temperature. The fact is that well-developed technologies in the neighboring infrared and millimeter wave bands are not directly applicable in the terahertz band. New materials, device physics, and integration techniques are required in the terahertz band. Great efforts are being made to fill the so-called 'terahertz gap' by maturing any existing terahertz device.

J. Sun, *Field-effect Self-mixing Terahertz Detectors*, Springer Theses,
DOI 10.1007/978-3-662-48681-8_1

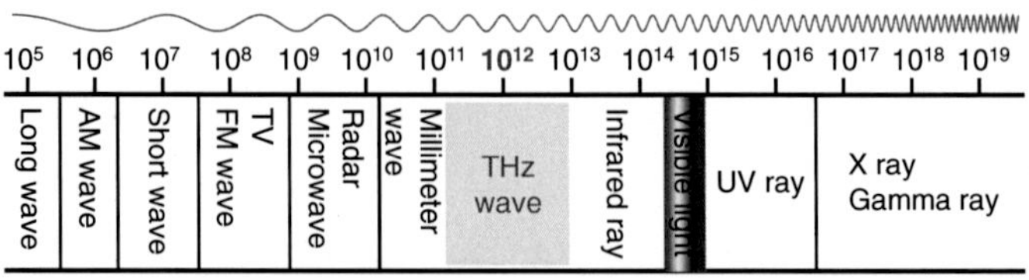

Fig. 1.1 The spectrum of electromagnetic wave form 10^5–10^{19} Hz.

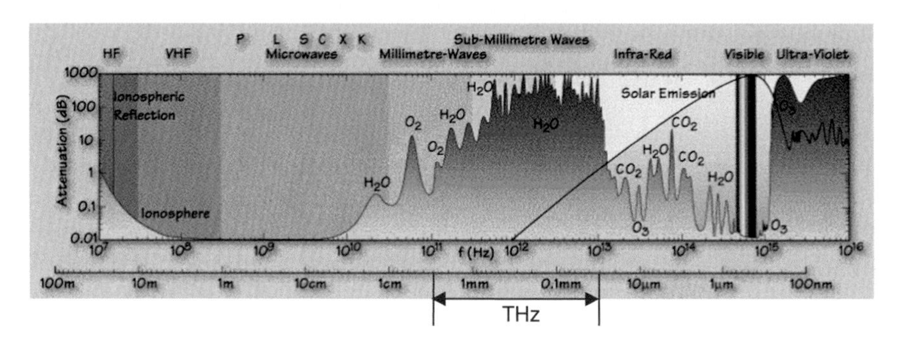

Fig. 1.2 The spectra of solar emission and the absorption in the earth atmosphere [5]

The earth's atmosphere is a strong absorber of terahertz radiation in specific water vapor absorption bands, so the range of terahertz radiation limits its applications in long-distance communications. However, at a distance of 10 m the band may still allow many applications in imaging and high bandwidth wireless networking, especially in indoor systems. In addition, producing and detecting coherent terahertz radiation remains technically challenging, though inexpensive commercial sources now exist in the 0.3–1.0 THz range (the lower part of the terahertz spectrum), including gyrotrons, backward wave oscillators, and resonant-tunneling diodes.

Therefore, terahertz waves have attracted the widespread attention of astronomers, who are gaining useful information about the universe such as the cosmic background radiation, which include the formation of stars, the composition of interstellar matter, and the key elements of the formation of life. The major atmospheric absorption features and the solar emissions are shown in Fig. 1.2. Some polar molecules in the atmosphere, such as water and carbon dioxide, have strong absorption of terahertz radiation. This limits the transmission distance and the practical application of terahertz waves.

Terahertz radiation is emitted as part of the black-body radiation from anything with temperatures greater than about 10 K. While this thermal emission is very weak, observations at these frequencies are important for characterizing the cold 10–20 K dust in the interstellar medium in the Milky Way galaxy and in distant starburst galaxies [1].

The long wave side is of electronics and the short wave side belongs to the category of photonics. Low frequency electromagnetic waves can be generated by the macroscopic movement of the charge. When the frequency achieves the terahertz band, the influence of the stray effects become more and more significant, which can be ignored at low frequency. Therefore, the traditional electromagnetic source and electronic components were not suitable for the terahertz band. We should develop faster and smaller size electronic components to meet the needs of the terahertz band. However, limited by the semiconductor processing techniques, it is difficult to fabricate smaller components. More importantly, the smaller components will be with some quantum effects, therefore, the classical electromagnetic wave theory will not be suitable. Corresponding to the photonics, the terahertz waves are derived from the quantum transition of two different energy levels. Because of the relatively low energy, the terahertz radiation is easily affected by thermal fluctuation. Therefore, these devices should usually work at low temperature. The basic theories of terahertz have been more mature, but limited by technology and the objective law, the terahertz applications need to be further improved.

Like infrared and microwave radiation, terahertz radiation travels in a line of sight and is nonionizing. Like microwave radiation, terahertz radiation can penetrate a wide variety of nonconducting materials, such as clothing, paper, cardboard, wood, masonry, plastic, and ceramics. The penetration depth is typically less than that of microwave radiation. Also, terahertz radiation has limited penetration through fog and clouds and cannot penetrate liquid water or metal [4].

With deeper understanding of terahertz waves, it has been found to possess many unique properties. Compared with other bands of electromagnetic waves, terahertz wave has the following characteristics and related applications:

1. **Rich molecule fingerprints and ionization-free**. Many molecular rotation/vibration modes and collective excitations in solids such as phonon and plasmon have specific fingerprints in the terahertz spectrum. Terahertz spectroscopy is being applied more and more widely in materials science and in identification of chemical substance such as drugs and hazardous and explosive substances. The terahertz photon energy of about 4.1 meV/THz is two orders of magnitude lower than that of X-ray and is insufficient to break strong covalent chemical bonds unlike X-ray. The terahertz wave provides a unique tool for exploring new molecular structures, molecular dynamics, and new medicines in life science. Various strong absorption lines in the terahertz band by water molecule may provide new methods to understand the unknown properties/functionalities of water and –OH in life. Such strong absorption, however, is unfavorable for applications in which a terahertz wave needs to be transmitted in any water-rich space.
2. **High spatial and temporal resolution**. Terahertz wave at 1 THz has a wavelength of 300 μm in free space. Compared to microwave and millimeter wave, the terahertz wave allows for a higher spatial resolution in imaging applications. The oscillation period in an order of 1 ps allows for accurate timing when the dynamics of molecule and solid is concerned. Terahertz time-domain spectroscopy revolutionizes the spectroscopy in terahertz frequency range. When optics and electronics meet in the terahertz regime, terahertz technology would ultimately allow for coherent control of molecular states and quantum states.

3. **High speed and high bandwidth**. The wide frequency range and high frequency characteristics of terahertz wave allow for high-speed and large-data communication/transmission for both wireless and on-chip communication. Information systems would leap into a new era when electronics and optoelectronics are merged in the terahertz band.
4. **High penetration**. Terahertz radiation can penetrate many kinds of nonmetallic and nonpolar materials include clothing, paper, cardboard, wood, masonry, plastic, ceramics, and etc. On the other hand, since it has a much longer wavelength than visible and infrared light, terahertz wave can be transmitted through smoke without severe scattering.

The unique nature of terahertz waves has stimulated intense research in terahertz science and technology. Terahertz technologies are believed to find various applications such as security screening [6–10], material composition analysis [11, 12], medical imaging [13–17], communication [18–20], nondestructive inspection [21, 22], astronomy observation [23–25], environmental monitoring [26] and etc. In 2005, the U.S. government identified the terahertz wave as one of the ten technologies that may change the future of the world and said "the twenty-first century is the terahertz era." Terahertz technology also topped the list of the 9 key technologies to develop in a research plan announced by the Japanese government in 2005.

1.2 Terahertz Sources

Terahertz radiation can be found from any object as long as the temperature is above absolute zero. However, this type of terahertz emission is of black-body type and the emission efficiency is low. Similar to millimeter-wave oscillators and light-emitting diodes/lasers, various electronic and optoelectronic approaches have been developed to increase and lower the oscillation frequency into the terahertz range, respectively. In the electronic approach, the devices include Schottky diode and nonlinear transmission lines for frequency multiplication, Esaki diodes and Gunn diodes for oscillators, and high-electron-mobility transistors (HEMTs) for oscillators and amplifiers. The electronic approach also includes vacuum electronic devices, such as traveling-wave tubes, gyrotrons, backward-wave oscillators (BWOs), and free-electron lasers for terahertz generation. The traveling-wave tubes are capable of terahertz amplification as well. In the optoelectronic approach, thermal emitters and lasers have been developed accordingly. Among these, the glow bar and the mercury lamp are well-known broadband terahertz sources widely used in Fourier-transform spectrometers (FTS). Terahertz quantum cascade lasers (THz-QCLs) are being developed into one of the most promising solid-state coherent terahertz sources. However, it is difficult for either electronic approach or optoelectronic approach to make practical terahertz sources that are compact, high efficiency & high power, and operatable at room temperature. As shown in Fig. 1.3, the well-known 'Terahertz Gap' refers to the region with both high power (1 mW–1 W) and high frequency (0.3–3 THz) which cannot

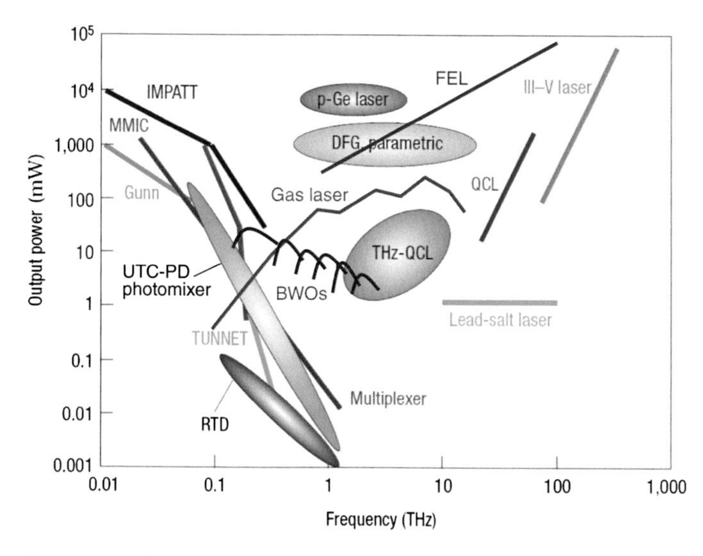

Fig. 1.3 Chart of emission power and frequency. *Solid lines* are for the conventional terahertz sources. IMPATT stands for impact ionization avalanche transit-time diode. MMIC stands for microwave monolithic integrated circuit. TUNNET stands for tunnel injection transit time device. Multiplexer stands for nonlinear devices such as SBD. Reprinted by permission of Macmillan Publishers Ltd Ref. [3], copyright 2007

be fully accessed with ease either by the electronic or the optoelectronic approach. Such inefficiency in both approaches has clear physical origins.

For electronics, the cut-off frequency of a device is determined by the electron transit time $\tau = L/v$, where L is the distance of electron transport and v is the drift velocity of the electron. To make a device at terahertz frequency, the transient time needs to be less than 1 ps, i.e., the device size has to be less than 100 nm and the electron drift velocity about 10^7 cm/s close to the saturation velocity in most of the semiconductors. Specifically in solid-state emitters, the parasitic circuit parameters such as series resistance and parasitic capacitance become indeed the limiting factors for cut-off frequency. Electrons in vacuum can be accelerated to relativistic speed and the device would be operated at terahertz frequencies when the dimension of the slow-wave structure is made into micrometer scale. Although the micromachining technique allows for fabrication of such miniatured vacuum electronic devices, difficulties lie in the realization of high-current-density electron beam for high-power operation. For optoelectronic devices, lower emission frequency means a smaller energy difference between two quantum states from which terahertz photons are emitted through quantum transition. The smaller the energy difference, the lower the operation temperature. The higher the temperature, the lower the efficiency.

Both electronic and photonic approaches have their own limitations in producing terahertz electromagnetic wave effectively. The marriage of both approaches would provide new opportunities to realize efficient generation of terahertz wave, as shown in Fig. 1.4. In fact, THz-TDS is a successful optical approach in combination with

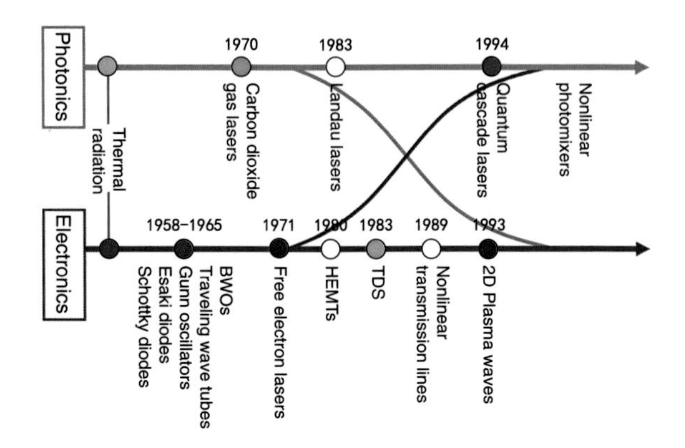

Fig. 1.4 The development of photonic and electronic terahertz sources

Table 1.1 A chart of the frequency range and output power for different terahertz sources

Terahertz sources	Frequency range (THz)	Output power (W)
Free electron lasers [31]	0.1–10	10^{1}–10^{2}
Optically pumped IR lasers [32]	0.1–8.0	10^{-1}–10^{-3}
Quantum cascade lasers [33]	1.2–4.5	10^{0}–10^{-3}
Gunn negative resistance oscillators [34]	0.03–0.3	10^{-1}–10^{-3}
IMMPATT diode oscillators [34]	0.03–0.4	10^{0}–10^{-4}
Backward wave oscillators [35]	0.2–1.1	10^{-2}–10^{-3}
Frequency multipliers [36]	0.1–1.7	10^{-1}–10^{-7}
Nonlinear photomixers [37]	0.1–1.6	10^{-5}–10^{-7}

the electronic approach. Photomixing and difference frequency generation (DFG), based on optoelectronic lasers utilize nonlinear electron transport in active optical medium, allows for generation of continuously tunable terahertz emission up to 5 THz [27, 28]. As another example, plasmon as collective oscillations of electron gas/liquid allows for generation and detection of terahertz electromagnetic wave [29]. Terahertz plasmonics involving intense coupling between terahertz electromagnetic wave and free electrons in either metallic or semiconductor structures have been intensively studied in a broad frequency range from visible to terahertz band.

As shown in Table 1.1, a comparison in terms of the frequency band and the output power of various continuous wave (CW) terahertz sources is given. The sources include free electron lasers (FELs), optically pumped infrared (IR) lasers, quantum cascade lasers (QCLs), Gunn and impact-ionization-avalanche-transit-time (IMPATT) diode oscillators, backward wave oscillators (BWOs), Schottky diode frequency multipliers, nonlinear photomixers, etc. Except FELs, the terahertz power delivered by most of the known terahertz sources is quite low, typically of the order of milliwatts and at most of the order of hundreds of milliwatts. The power of terahertz waves should be higher than 1 mW for most applications. For security applications, likely terahertz power between 1 mW and 1 W is required, while in medical imaging applications, irradiation power is restricted to 1 $\mathrm{mW/cm^2}$, limiting the maximum imaging depth due to attenuation of the terahertz signal within 1 mm [30]. Figure 1.4 illustrates the source development. There are three major approaches for developing the terahertz source [3]. The first is optical terahertz generation, which has spearheaded terahertz research in the past few decades. The THz-QCL is under development. The second uses solid-state electronic devices,which are already well established at low frequencies. There are lack of compact, portable, low cost, strong power, and commercial terahertz sources at the center of 1 THz.

1. **Free Electron Lasers (FELs)** offer high output power of the order of hundreds of watts to, potentially, kilowatts. In FELs, free electrons in vacuum are accelerated to relativistic speed, then run through a 'wiggler' where they give up energy, which is converted into light. Its frequency can be adjusted continuously within 0.1–10 THz. Because of the complex structure, huge system, and high cost, FELs are only built in a few research institutions and are mainly used for scientific research.
2. **Backward Wave Oscillators (BWOs)**, also known as Carcinotrons, are a type of traveling wave tube, where energy from an electron beam traveling the length of an evacuated tube in a high magnetic field is absorbed into a waveguide intersecting the tube at regular intervals. The output frequenciey can be tuned by the electron beam accelerating potential from 0.1–1 THz.
3. **Gunn and IMPATT Diodes** are wildly used in many microwave oscillators. These diodes provide a peculiar negative differential conductance (NDC) in the high frequency band. Implemented in an oscillatory circuit/cavity, small fluctuations in the bias voltage tend to grow and result in self-oscillations. Simulations of GaN IMPATT diodes operating in the frequency range up to 0.7 THz have been reported recently [38]. Although oscillators based on Gunn diode or IMPATT diodes are not yet able to work at terahertz frequencies, Schottky barrier diodes can be used as frequency multipliers to generate terahertz emission [39]. On the other hand, resonant tunneling diodes (RTDs) can provide negative differential conductance at terahertz frequencies. Terahertz oscillators based on RTDs embedded in a slot antenna have been demonstrated with a frequency above 1 THz at room temperature [40–42]. The common drawback of the NDC-based approach and frequency multiplication is the reduction in emitted power by increasing the frequency.
4. **Terahertz Gas Lasers** typically consist of a low pressure molecular gas laser pumped by a grating tuned CO_2 optical laser [43]. The output power ranges from milliwatts to hundreds of milliwatts over a range of several THz. Due to the discrete quantum energy levels of gas molecules, the output frequency distributes only in a few limited terahertz frequencies and cannot be continuously adjusted. Since this type of laser is usually bulky, high power consumption, terahertz gas lasers are generally used in laboratories.
5. **Quantum Cascade Lasers (QCLs)** consist of a stack of repeated quantum well structures (typically 20–200) in which photons are created via intersubband transitions within the conduction band. QCLs have high quantum efficiency due to the cascading effect. The output power is typically in the range of tens to hundreds

of milliwatts [44, 45]. The output frequencies can be adjusted using a series of different quantum well structures from 1.2 to 4.5 THz [46]. While not yet capable of room temperature operation, recent devices have been demonstrated at 225 K [47].

6. **Photomixers** typically consist of an inter-digitized antenna structure across which a DC bias is applied. Terahertz radiation at the difference frequency of two incident optical lasers arises from the photoelectric effect. The radiation is coupled to free space using a planar antenna and/or a hemispherical silicon lens [48]. Although the efficiency is low in the range of 10^{-6}–10^{-3}, high power terahertz pulses may be generated since the optical pump lasers could deliver very high input power.

1.3 Terahertz Detectors

Terahertz detectors are one type of key components for terahertz applications in which generation of electrical signal is induced by a weak incident terahertz electromagnetic wave. Without the need to deal with high electrical power, terahertz detectors are relatively easier to be made. However, similar physical limitations found in terahertz emitters are then encountered in terahertz detectors. It is still demanding to develop terahertz detectors with precedent sensitivity at high frequencies and especially for room temperature applications. Different detection mechanisms can be similarly divided into two categories, i.e., photonic and electronic.

Photonic detectors include terahertz quantum-well photodetectors (THz-QWP) [49] and bolometric detectors, both relying on the absorption of terahertz photons in spite of the different facts that electronic states or lattice vibration modes are excited, respectively. For photonic detectors, the sensitivity is limited by the level of thermal fluctuation and ultimately photo background noise. The lower the temperature, the smaller the background noise and the higher the sensitivity. Well-known bolometric detectors are liquid-helium-cooled silicon bolometers and superconducting hot-electron bolometers for low-temperature applications, pyroelectric detector and VO_x-based microbolomter array for room temperature applications. The speed of a bolometric detector is determined by the heat capacity and the thermal conductance. At cryogenic temperatures where the number of background photons can be greatly suppressed, bolometric detectors can have both high sensitivity and high speed. For those bolometric detectors operated at room temperature, small heat capacity, low thermal conductance, and hence slow speed are required to maintain a significant signal-to-noise ratio with a large thermal background. The low speed is usually the main drawback of room-temperature bolometric detectors, which are not suitable for high-speed applications.

Electronic detectors sense the alternating terahertz electric field and generate a corresponding electrical signal at low frequency. Hence, electronic detectors usually have nonlinear current-voltage characteristics and are called rectifiers or mixers. Terahertz rectifiers are of direct detection type, which senses the amplitude of the incident power or the amplitude of the terahertz electric field. A terahertz mixer usually mixes the terahertz signal with a strong CW terahertz signal (called local signal) and generates an electrical signal at an intermediate frequency, which is the frequency difference between the two terahertz signals. Terahertz mixers allow for heterodyne detection of not only the amplitude but also the phase of the terahertz signal. In comparison, direct detectors can be called homodyne/self-mixing detectors that do not require a local terahertz source.

With the advent of femtosecond lasers, electrooptical sampling technique allows for a new type of terahertz detection and spectroscopy. This technique is now widely implemented in THz-TDS systems. The sampling is realized in the way that a continuously time delayed femto-second laser pulse is used to probe the instant terahertz electric field in the time domain. The optical probe beam and the terahertz beam are focused onto a nonlinear electrooptical crystal in which the polarization of the probe beam is altered by the terahertz electric field.

High sensitivity in detection is highly desirable, especially when the terahertz signal is relatively weak. The responsivity characterizes the ability of a detector in transducing the incident terahertz power (in Watt) into the electrical signal (in Ampere or Volt). The sensitivity is the most important figure of merit for characterizing the minimum terahertz power which can be sensed and transduced into a measurable electrical signal without being immersed in the noise background. Noise equivalent power (*NEP*, in $W/\sqrt{Hz}$), defined as the ratio of the spectral density of the electrical noise (in $A/\sqrt{Hz}$ or $V/\sqrt{Hz}$) to the responsivity (in A/W or V/W), is the most commonly applied for sensitivity characterization.

Except for the responsivity and sensitivity, spectral response, speed, and operation temperature are also important quantities when specific sensing applications are concerned. As shown in Table 1.2, detectors operated at cryogenic temperatures and room temperature are listed for comparison. Cryogenic detectors include silicon bolometers, hot-electron bolometer (HEBs), superconducting tunnel junction

Table 1.2 Different characteristics of various types of terahertz detectors

Terahertz detectors	NEP ($W/Hz^{1/2}$)	Responsivity (V/W)	Resp. time (s)	Op. Temp (K)	Freq. range (THz)
Cryogenic detectors					
Bolometers [50]	10^{-16}–10^{-13}	10^7–10^5	10^{-2}–10^{-3}	≤ 4.2	0.1—30
HEB [51, 52]	10^{-19}–10^{-17}	10^9	10^{-8}	≤ 0.3	0.1–30
STJ[a] [53]	10^{-16}	10^9	10^{-3}	≤ 0.8	0.1–1
Room temp. detectors					
Golay cells [54, 55]	10^{-10}–10^{-8}	10^5–10^4	10^{-2}	300	0.1–20
Pyroelectrics [56]	10^{-9}	10^5	10^{-2}	240–350	0.1–30
Schottky diodes [57]	10^{-12}	10^3	10^{-11}	10–420	0.1–1.7
HEMTs [58]	10^{-11}	10^3	10^{-11} [59]	4.2–420	0.1–4
Silicon FETs [60, 61]	10^{-10}	10^2	10^{-9} [59]	10–420	0.1–4.3

[a]Superconducting tunnel junction (STJ)

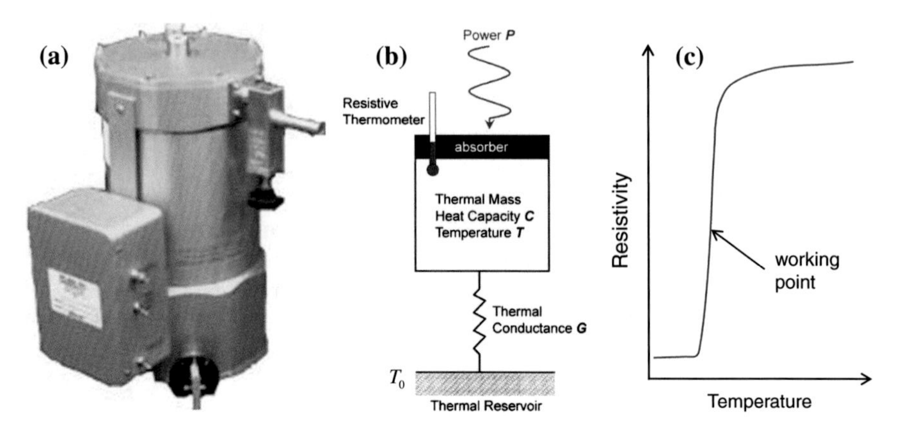

Fig. 1.5 **a** The photo of the bolometer. **b** Schematic diagram. **c** The resistivity of the thermistor as a function of the temperature

(STJs), etc. Room temperature detectors include Golay cells, pyroelectric detectors, Schottky-barrier diodes (SBDs), HEMTs, silicon field-effect transistors (FETs), etc.

1. **Bolometers** rely on a temperature-dependent resistor whose resistance can be sensitively altered by the minute power of an incident electromagnetic radiation [62, 63]. As an example, a liquid-heilium cooled silicon bolometer and the sensing mechanism are shown in Fig. 1.5a and b, respectively. The key element is a thin silicon bridge whose resistance is a sensitive function of the temperature as shown in Fig. 1.5c. To convert the energy of the incident electromagnetic wave into heat sensed by the silicon bridge, an absorber which is usually made of a thin layer of metal, is attached to the silicon bridge. Any radiation impinging on the absorptive element raises the bridge's temperature above that of the thermal reservoir (the liquid-heilium temperature). The temperature change is proportional to the incident power and is reversely proportional to the heat capacity of the silicon bridge (including that of the absorber if there is any) and the thermal conductance between the bridge and the thermal reservoir. A minute temperature change can be detected by measuring the varied resistance. The intrinsic thermal time constant, which sets the speed of the bolometer, is equal to the ratio of the heat capacity to the thermal conductance.
   Bolometers use semiconductor or superconductor elements, enabling significantly greater sensitivity; they are usually operated at cryogenic temperatures (below 10 K). Therefore, compared to more conventional particle detectors, they are extremely efficient in energy resolution and in sensitivity. The NEP is usually lower than $1 \times 10^{-13}\ \text{W}/\sqrt{\text{Hz}}$ and the responsivity is usually higher than $1 \times 10^{7}\ \text{V/W}$. Because bolometers are thermal detectors, the frequency range is broad at 0.1–30 THz. Bolometers are usually very slow to reset (about 10^{-2}–10^{-3} s), therefore, they are not used for high-speed imaging and communication. In addition, cryogenic bolometers are bulky.

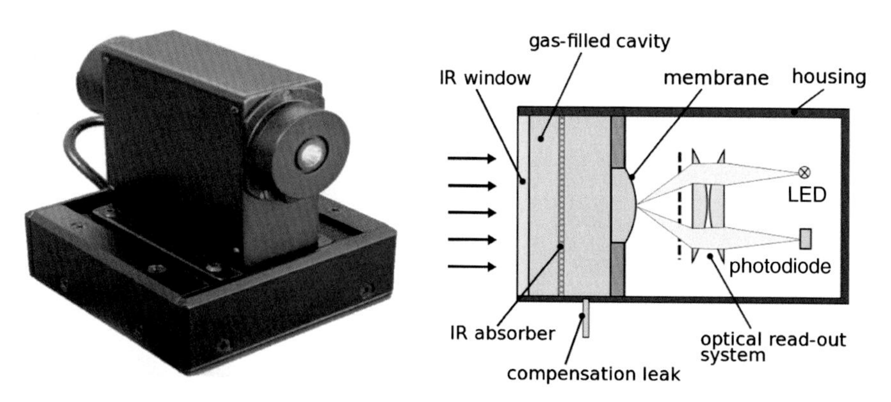

Fig. 1.6 The photo and the schematic diagram of the Golay cell detector [54]

2. **Golay Cell** is indeed another type of bolometric detector frequently used for sensitive terahertz sensing. The detection of heat is realized by using the opto-acoustic effect. The concept was originally described by Golay [64], after whom the Golay cell was named. An actual Golay cell and its schematic diagram are given in Fig. 1.6. In the Golay cell, a gas-filled cavity with a flexible diaphragm or membrane and a window for receiving the incident terahertz wave serves as the sensing element. An absorber fixed inside the gas cavity absorbs the incident terahertz wave and heats up the gas in the cavity. With increasing temperature, the gas expands and the membrane is deformed accordingly. The deformation is a direct measure of the incident energy absorbed by the gas cavity and can be sensed by detecting the deflected light from the membrane, as shown in Fig. 1.6. Golay cells have a relatively high sensitivity at room temperature and a flat response over a very broad range of frequencies from 0.1 to 30 THz. The responsivity and the *NEP* are about 1×10^5 V/W and 1×10^{-10} W/$\sqrt{\text{Hz}}$, respectively. The speed of a Golay cell depends on the heat capacity of the cavity and the heat conductance. Golay cells are slow detectors with a response time of the order of a few ten milliseconds. Golay cells are susceptible to mechanical vibrations.
3. **Pyroelectric Detector** is bolometric and relies on the thermal pyroelectric effect. A photo and schematic diagram of a pyroelectric detector are given in Fig. 1.7. The key element of the detector is a thin pyroelectric film, which is in the form of a capacitor. To receive the incident terahertz electromagnetic wave, an absorptive film is deposited on one of the plates of the capacitor. Upon incident teraheratz wave, the elevated temperature induces a polarization change and hences the capacitance. By measuring the capacitance or the induced charging/discharging current when it is biased at a constant bias voltage, the incident power can be deduced. It has to be noted that only modulated radiation creates a signal, therefore either pulsed or mechanically chopped terahertz sources are used and the background radiation is filtered out. The response time is slower than 10 ms, therefore, the modulation frequency should be lower than 100 Hz. Since pyroelectric detectors operate on a thermal phenomenon, they have a very broad spectral

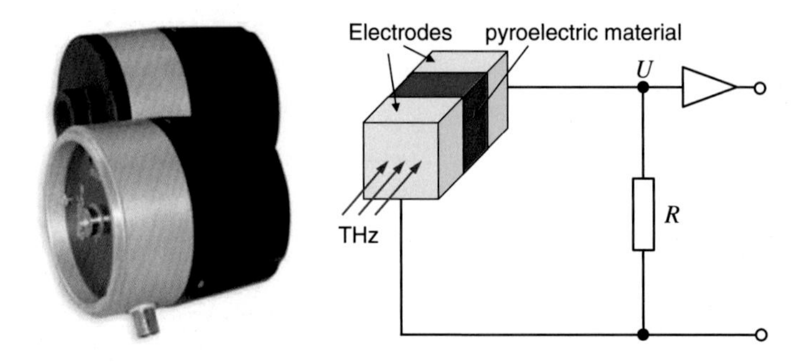

Fig. 1.7 The photo and the schematic diagram of the pyroelectric detector [56]

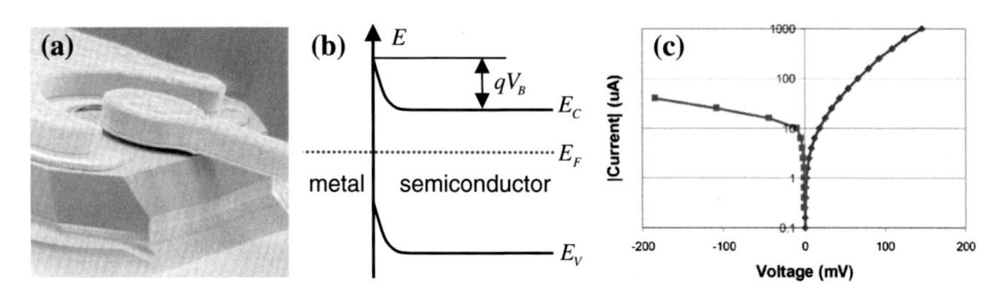

Fig. 1.8 **a** Optical microscope image of a Schottky diode produced by Virginia Diode Inc. **b** The energy band structure of the Schottky barrier. **c** Typical $I - V$ characteristics [65]

response from 0.1 to 30 THz without any cooling like semiconductor detectors. The responsivity is usually higher than 1×10^5 V/W, and the NEP can be as low as 4×10^{-10} W/$\sqrt{\text{Hz}}$ at 300 K.

4. **Schottky-Barrier-Diode Detectors** rely on the nonlinear $I - V$ characteristic, which rectifies the incident terahertz electromagnetic wave. A photo and a schematic of a GaAs Schottky barrier diode (SBD) produced by Virginia Diode Inc. are shown in Fig. 1.8. The Schottky junction is formed between the metallic anode and the n-type semiconductor as the cathode. As shown in Fig. 1.8c, a zero-bias SBD allows for sensitive detection of terahertz wave. The cut-off frequency of a SBD is determined by its junction resistance and the capacitance. To reduce the capacitance, the effect of the high dielectric constant of GaAs is minimized by making the metallic anode in a form of air bridge and thinning/replacing the GaAs substrate with a low dielectric substrate such as quartz. The resistance of the diode is minimized by improving the Schottky barrier. The cut-off frequency of a typical GaAs SBD detector is more than 1 THz. In a SBD detector, the diode is integrated in a terahertz wave guide or on a hyperhemispheric high-resistivity silicon lens. SBD detectors provide high sensitivity of the order of 10 pW/$\sqrt{\text{Hz}}$ for frequency below 0.7 THz; and the corresponding responsivity is of the order of 10^3 V/W. Above 0.7 THz, the responsivity drops to the order of 10^2 V/W and the sensitivity is about 40 pW/$\sqrt{\text{Hz}}$.

5. **Field-Effect Terahertz Detector** is a new type of solid-state electronic/plasmonic terahertz detector. Terahertz detection in FETs was first proposed by M.I. Dyakonov and M.S. Shur from the point of view of plasma wave [66, 67]. As a collective mode of dense electron gas, plasma waves could render a new approach for the realization of room-temperature terahertz detection. The schematic diagram and the principal diagram of the field-effect terahertz detector are shown in Fig. 1.9. In hydrodynamic approximation, the plasmon mode frequency in a two-dimensional electron gas (2DEG) can be expressed as [66]

$$\omega_0 = \frac{\pi}{2L}\sqrt{\frac{e(V_g - V_{th})}{m^*}}, \tag{1.1}$$

where L is the gate length, V_g is the gate voltage, V_{th} is threshold voltage, and m^* is the effective electron mass. When a terahertz radiation is fed into the electron channel between the gate and the source, the carrier density and the carriers drift velocity under the gate are simultaneously modulated by the incident terahertz field. As a result, homodyne mixing (self-mixing) of the terahertz field in the gated electron channel induces a DC photocurrent. Such a homodyne mixing effect occurs when plasma waves are resonantly or nonresonantly excited in the gated channel. For a field-effect channel with high electron mobility, e.g., in III–V heterostructures or at cryogenic temperatures, plasma waves excited can propagate through the channel from the source to the drain leading to resonant detection of plasma waves. In silicon-based field-effect channel and others with relatively low carrier mobility, plasma waves excited are usually overdamped and is effective only in a small portion of the gated electron channel. Since field-effect plasmon detectors rely on the collective oscillations of electron gas/liquid, parasitic capacitance of the gated electron channel has less effect on the cut-off frequency and the sensitivity. FET-based homodyne mixing is effective in the frequency range from a few gigahertz to 5 THz [61]. The responsivity and the NEP are expected to surpass those of the SBD. The readout speed is limited by the cut-off frequency of the FET and can be of the order of a few ten gigahertz [59]. Hence, FET-based plasmon detectors are expected to play an important role in terahertz applications.

Through the above comparison of various terahertz detectors, bolometric detectors have broadband response and semiconductor detectors are usually narrow-band, but have high response speed. Cryogenic bolometers have the highest detection sensitivity and the NEP is in the range of 10^{-13}–10^{-16} $\mathrm{W}/\sqrt{\mathrm{Hz}}$. Pyroelectric detectors and the Golay cells offer high sensitivity at room temperature, however, they are slow in response. Field-effect detectors as will be shown in this thesis may provide high sensitivity, broad-band response, and high speed at room temperature. FET detectors can be made by using standard fabrication techniques without the need for making air-bridge electrodes. Furthermore, FET detectors are able to sense terahertz wave with frequency well above 1 THz [61]. FET detectors have great potential to replace SBD detectors in many applications.

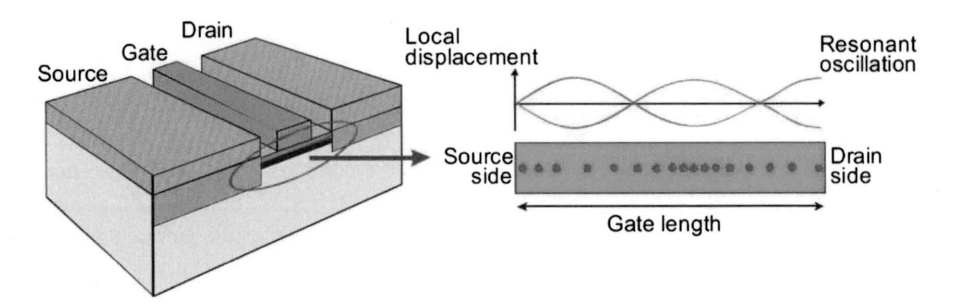

Fig. 1.9 The schematic diagram and the principle diagram of the field-effect terahertz detector

1.4 Outline of This Thesis

Terahertz science and technology are receiving great attention due to the peculiar effects and wide application prospects. The lack of mature technologies for manufacturing terahertz sources and detectors become the bottleneck in the development of this field. Field-effect detectors are expected to achieve these features. Currently, field-effect detectors are in the initial stage of development. Because AlGaN/GaN HEMTs have excellent electrical characteristics, they have become the alternative to high performance terahertz detectors. In the following six chapters, We will focus on the principle, design, fabrication, characterization, and application of the AlGaN/GaN detectors [58, 68–72]. In this thesis, seven chapters are prepared and arranged in the following way.

In this chapter, we introduce the background and the features of terahertz waves. Then, we briefly present the principle and characteristics of several representative terahertz sources. Finally, we focus on several commonly used terahertz detectors, their advantages, and disadvantages.

In Chap. 2, we summarize the existing theories on terahertz plasmon detection. We introduce a quasi-static self-mixing detection model that takes into account the localized terahertz fields. The model can extend the detection theory to all working regimes of field-effect transistors. Based on this theory, we propose a series of antenna design schemes and simulate the antennas by the finite-difference time domain method. Finally, we briefly introduce the terahertz low-pass filter for antenna isolation.

In Chap. 3, we introduce experiments of the self-mixing detectors based on antenna coupled AlGaN/GaN HEMTs. First, we briefly introduce the AlGaN/GaN heterojunction and their advantages. We introduce the typical structure and basic characteristics of the AlGaN/GaN HEMTs. We also introduce the fabrication process of the terahertz detectors in details. Then, we study AlGaN/GaN detectors without and with source–drain bias. We focus on the basic properties of the detectors, including the responsivity, NEP, response speed, response frequency, and polarization. Finally, we give the comparison of a series of detectors with different terahertz antennas.

In Chap. 4, we demonstrate and analyze the resonant plasmon detection in an AlGaN/GaN 2DEG detector with symmetrically arranged nanogates, antennas, and filters.

In Chap. 5, we perform a near-field imaging experiment on an antenna-coupled field-effect terahertz detector. The combined use of a scanning metallic probe and the field-effect terahertz detector allows us to image the active region of terahertz antennas. This method provides an alternative way for rapid verification of a given terahertz antenna design.

In Chap. 6, we demonstrate some terahertz applications based on field-effect terahertz detectors including the terahertz imaging and the Fourier transform spectroscopy etc.

In Chap. 7, we summarize the findings of this thesis and present an outlook for future studies.

References

1. Introduction of terahertz radiation from Wikipedia (2014). http://en.wikipedia.org/wiki/Terahertz_radiation. Accessed 17 Sept 2014
2. Zhang, X.C., Xu, J.Z.: Introduction to THz Wave Photonics. Springer, Heidelberg (2010)
3. Tonnuchi, M.: Cutting-edge terahertz technology. Nat. Photonics **1**, 97–105 (2007)
4. JLab generates high-power terahertz light (2014). http://cerncourier.com/cws/article/cern/28777. Accessed 10 Jan 2014
5. The electromagnetic spectrum (2014). http://envisat.esa.int/support-docs/em-spectrum/em-spectrum.html. Accessed 25 March 2014
6. Chen, Y.Q., Liu, H.B., Deng, Y.Q., Schauki, D., Fitch, M.J., Osiander, R., Dodson, C., Spicer, J.B., Shur, M., Zhang, X.C.: THz spectroscopic investigation of 2,4-dinitrotoluene. Chem. Phys. Lett. **400**, 357–361 (2004)
7. Shen, Y.C., Lo, T., Taday, P.F., Cole, B.E., Tribe, W.R., Kemp, M.C.: Detection and identification of explosives using terahertz pulsed spectroscopic imaging. Appl. Phys. Lett. **86**, 24116 (2005)
8. Liu, H.B., Chen, Y.Q., Bastiaans, G.J., Zhang, X.C.: Detection and identification of explosive RDX by THz diffuse reflection spectroscopy. Opt. Express **14**(1), 415–423 (2006)
9. Funk, D.J., Calgaro, F., Averitt, R.D., Asaki, M.L.T., Taylor., A.J.: THz transmission spectroscopy and imaging: application to the energetic materials PBX 9501 and PBX 9502. Appl. Spectrosc. **58**(4), 428–431 (2004)
10. Kemp, M.C., Taday, P.F., Cole, B.E., Cluff, J.A., Fitzgerald, A.J., Tribe, W.R.: Security applications of terahertz technology. Proc. SPIE **5070**, 44 (2003)
11. Choi, M.K., Bettermann, A., van der Wiede, D.W.: Potential for detection of explosive and biological hazards with electronic terahertz systems. Philosophical transactions. Math. Phys. Eng. Sci. **362**, 337 (2004)
12. Fergusona, B., Wanga, S., Zhong, H., Abbottc, D., Zhanga, X.C.: Powder retection with T-ray imaging. Proc. SPIE **5070**, 7 (2003)
13. Arnone, D.D., Ciesla, C.M., Corchia, A., Egusa, S., Pepper, M., Chamberlain, J.M., Bezant, C., Linfield, E.H.: Application of terahertz (THz) technology to medical imaging. Proc. SPIE **3828**, 209 (1999)
14. Cho, G.C., Han, P.Y., Zhang, X.C.: Time-domain transillumination of biological tissues with terahertz pulses. Opt. Lett. **25**(4), 242–244 (2000)
15. Mickan, S.P., Menikh, A., Liu, H., Mannella, C.A., MacColl, R., Abbott, D., Munch, J.: Label-free bioafnity detection using terahertz technology. Phys. Med. Biol. **47**, 3789 (2002)

16. Alexandrov, B.S., Gelev, V., Bishop, A.R., Usheva, A., Rasmussen, K.O.: DNA breathing dynamics in the presence of a terahertz field. Phys. Lett. A **374**(10), 1214–1217 (2010)
17. Swanson, E.S.: Modelling DNA response to THz radiation. Phys. Rev. E **83**(4), 040901 (2010)
18. Kleine-Ostmann, T., Nagatsuma, T.: A review on terahertz communications research. J. Infrared Millimeter Terahz Waves **32**(2), 143–171 (2011)
19. Jha, K.R., Singh, G.: Terahertz Planar Antennas for Next Generation Communication. Springer, Heidelberg (2014)
20. Ishigaki, K., Shiraishi, M., Suzuki, S., Asada, M., Nishiyama, N., Arai, S.: Direct intensity modulation and wireless data transmission characteristics of terahertz-oscillating resonant tunnelling diodes. Electron. Lett. **48**(10), 582–583 (2012)
21. Zhong, H., Xu, J.Z., Xie, X., Yuan, T., Reightler, R., Madaras, E., Zhang, X.C.: Nondestructive defect identification with terahertz time-of-flight tomography. IEEE Sens. J. **5**(2), 203–208 (2005)
22. Sanchez, A.R., Karpowicz, N., Xu, J.Z., Zhang, X.C.: Damage and defect inspection with terahertz waves. In: Proceedings of the 4th International Workshop on Ultrasonic and Advanced Methods for Nondestructive Testing and Material Characterization, vol. 4, p. 67 (2006)
23. Benford, D.J., Amato, M.J., Mather, J.C., Moseley Jr, S.H., Leisawitz, D.T.: Mission concept for the single aperture far-infrared (safir) observatory. Appl. Sp. Sci. **294**(3), 177–212 (2004)
24. Siegel, P.H.: THz applications for outer and inner space. In: Proceedings of the 17th International Zurich Symposium on Electromagnetic Compatibility, vol. 17, p. 1 (2006)
25. Solomon, S.: The mystery of the antarctic ozone "hole". Rev. Geophys. **26**(1), 131–148 (1988)
26. Pereira, M.F., Shulika, O.: Terahertz and Mid Infrared Radiation: Detection of Explosives and CBRN (using terahertz), pp. 153–165. Springer, Heidelberg (2014)
27. Vijayraghavan, K., Jiang, Y., Jang, M., Jiang, A., Choutagunta, K., Vizbaras, A., Demmerle, F., Boehm, G., Amann, M.C., Belkin, M.A.: Broadly tunable terahertz generation in mid-infrared quantum cascade lasers. Nat. Commun. **4**, 2021 (2013)
28. Ito, H., Nakajima, F., Furuta, T., Ishibashi, T.: Continuous THz-wave generation using antenna-integrated uni-travelling photodiodes. Semicond. Sci. Technol. **20**, S191–S198 (2005)
29. Popov, V.V.: Plasmon excitation and plasmonic detection of terahertz radiation in the grating-gate field-effect-transistor Structures. J. Infrared Millimeter Terahz Waves **32**, 1178–1191 (2011)
30. Salmon, N.A.: Scene simulation for passive and active millimetre and submillimetre wave imaging for security scanning and medical applications. Proc. SPIE **5619**, 129 (2004)
31. The FEL program at Jefferson Lab (2014). https://www.jlab.org/FEL/. Accessed 6 Jan 2014
32. Inguscio, M., Evenson, K.M., Moruzzi, G., Jennings, D.A.: A review of frequency measurements of optically pumped lasers from 0.1 to 8 THz. J. Appl. Phys. **60**, R161 (1986)
33. Williams, B.S.: Terahertz quantum-cascade lasers. Nat. Photonics **1**, 517 (2007)
34. Haddad, G.I., East, J.R., Eisele, H.: Two-terminal active devices for terahertz sources. Int. J. High Speed Electron. Syst. **13**, 395 (2003)
35. Tretyakov, M.Y., Volokhov, S.A., Golubyatnikov, G.Y., Karyakin, E.N., Krupnov, A.F.: Compact tunable radiation source at 180–1500 GHz frequency range. Int. J. Infrared Millimeter Waves **20**(8), 1443–1451 (1999)
36. Crowe, T.W., Porterfield, D.W., Hesler, J.L., Bishop, W.L., Kurtz, D.S., Kai, H.: Terahertz sources and detectors. Proc. SPIE **5790**, 271 (2005)
37. Mikulics, M., Schieder, R., Michael, E.A., Stutzki, J., Gusten, R., Marso, M., van der Hart, A., Bochem, H.P., Luth, H., Kordos, P.: Traveling-wave photomixer with recessed interdigitated contacts on low-temperature-grown GaAs. Appl. Phys. Lett. **88**, 041118 (2006)
38. Reklaitis, A., Reggiani, L.: Monte Carlo study of hot-carrier transport in bulk wurtzite GaN and modeling of a near-terahertz impact avalanche transit time diode. J. Appl. Phys. **95**, 7925 (2004)
39. Ward, J., Schlecht, E., Chattopadhyay, G., Maestrini, A., Gill, J., Maiwald, F., Javadi, H., Mehdi, I.: Capability of THz sources based on Schottky diode frequency multiplier chains. In: 2004 IEEE MTT-S International Microwave Symposium Digest, vol. 3, pp. 1587–1590 (2004)

40. Suzuki, S., Asada, M., Teranishi, A., Sugiyama, H., Yokoyama, H.: Fundamental oscillation of resonant tunneling diodes above 1 THz at room temperature. Appl. Phys. Lett. **97**, 242102 (2010)
41. Feiginov, M., Sydlo, C., Cojocari, O., Meissner, P.: Resonant-tunnelling-diode oscillators operating at frequencies above 1.1 THz. Appl. Phys. Lett. **99**, 233506 (2011)
42. Britnell, L., Gorbachev, R.V., Geim, A.K., Ponomarenko, L.A., Mishchenko, A., Greenaway, M.T., Fromhold, T.M., Novoselov, K.S., Eaves, L.: Resonant tunnelling and negative differential conductance in graphene transistors. Nat. Commun. **4**, 1794 (2013)
43. Terahertz gas lasers of Coherent, Inc (2012). http://www.cohr.com/downloads/opticallypumpedlaser.pdf. Accessed 10 March 2012
44. Kohler, R., Tredicucci, A., Beltram, F., Beere, H.E., Linfield, E.H., Davies, A.G., Ritchie, D.A., Dhillon, S.S., Sirtori, C.: High-performance continuous-wave operation of superlattice terahertz quantum-cascade lasers. Appl. Phys. Lett. **82**, 1518 (2003)
45. Williams, B.S., Kumar, S., Qin, Q., Hu, Q., Reno, J.L.: High-temperature and high-power terahertz quantum-cascade lasers. Proc. SPIE **6485**, 64850M (2007)
46. Walther, C., Fischer, M., Scalari, G., Terazzi, R., Hoyler, N., Faist, J.: Quantum cascade lasers operating from 1.2 to 1.6 THz. Appl. Phys. Lett. **91**, 131122 (2007)
47. Wade, A., Fedorov, G., Smirnov, D., Kumar, S., Williams, B.S., Hu, Q., Reno, J.L.: Magnetic-field-assisted terahertz quantum cascade laser operating up to 225 K. Nat. Photonics **3**, 41 (2009)
48. Brown, E.R.: THz Generation by photomixing in ultrafast photoconductors. Int. J. High Speed Electron. Syst. **13**, 497 (2003)
49. Yu, C.H., Zhang, B., Lu, W., Shen, S.C., Liu, H.C., Fang, Y.Y., Dai, J.N., Chen, C.Q.: Strong enhancement of terahertz response in GaAs/AlGaAs quantum well photodetector by magnetic field. Appl. Phys. Lett. **97**, 022102 (2010)
50. Richards, P.L.: Bolometers for infrared and millimeter waves. J. Appl. Phys. **76**, 1 (1994)
51. Kuzmin, L., Fominsky, M., Kalabukhov, A., Golubey, D., Tarasov, M.: Capacitively coupled hot-electron nanobolometer with SIN tunnel junctions. Proc. SPIE **4855**, 217 (2002)
52. Stevenson, T.R., Hsieh, W.T., Mitchell, R.R., Isenberg, H.D., Stahle, C.M., Cao, N.T., Schneider, G., Travers, D.E. Harvey M.S., Wollack, E.J., Henry, R.M.: Silicon hot-electron bolometers with single-electron transistor readout. Nucl. Instrum. Methods Phys. Res., Sect. A **559**, 591 (2006)
53. Ariyoshi, S., Otani, C., Dobroiu, A., Sato, H., Kawase, K., Shimizu, H.M., Taino, T., Matsuo, H.: Terahertz imaging with a direct detector based on superconducting tunnel junctions. Appl. Phys. Lett. **88**, 203503 (2006)
54. Golay cells by MICROTECH instuments, Inc (2014). http://www.mtinstruments.com/thzdetectors/index.htm. Accessed 12 May 2014
55. Golay cells by TYDEX (2014). http://www.tydexoptics.com/en/products/thz_optics/golay_cell/. Accessed 11 Apr 2014
56. Pyroelectric detectors by gentec-eo (2014). http://www.spectrumdetector.com/. Accessed17 Jun 2014
57. Chahal, P., Morris, F., Frazier, G.: Zero bias resonant tunnel Schottky contact diode for wide-band direct detection. IEEE Electron Device Lett. **26**(12), 894–896 (2005)
58. Sun, J.D., Sun, Y.F., Wu, D.M., Cai, Y., Qin, H., Zhang, B.S.: High-responsivity, low-noise, room-temperature, self-mixing terahertz detector realized using floating antennas on a GaN-based field-effect transistor. Appl. Phys. Lett. **100**, 013506 (2012)
59. Kachorovskii, V.Y., Shur, M.S.: Field effect transistor as ultrafast detector of modulated terahertz radiation. Solid State Electron. **52**(2), 182–185 (2008)
60. Tauk, R., Teppe, F., Boubanga, S., Coquillat, D., Knap, W.: Plasma wave detection of terahertz radiation by silicon field effects transistors: responsivity and noise equivalent power. Appl. Phys. Lett. **89**, 253511 (2006)
61. Boppel, S., Lisauskas, A., Mundt, M., Seliuta, D., Minkevičius, L., Kašalynas, I., Valušis, G., Mittendorff, M., Winnerl, S., Krozer, V., Roskos, H.G.: CMOS integrated antenna-coupled field-effect transistors for the detection of radiation from 0.2 to 4.3 THz. IEEE Trans. Microwave Theory Tech. **60**, 3834 (2012)

62. Bolometers by Infrared Laboratories, Inc (2014). http://www.infraredlaboratories.com/Bolometers.html. Accessed 27 Jun 2014
63. Zhou, J.W., Farooqui, K., Timbie, P.T., Wilson, G.W., Allen, C.A., Moseley, S.H., Mott, D.B.: Monolithic silicon bolometers as sensitive MM-wave detectors. IEEEMTT-S Int. Microwave Symp. Digest **3**, 1347 (1995)
64. Golay, M.J.E.: Theoretical consideration in heat and infra-red detection, with particular reference to the pneumatic detector. Rev. Sci. Instrum. **18**, 347 (1947)
65. Schottky-barrier-diode detectors by Virginia Diodes, Inc (2015). http://www.vadiodes.com/index.php/en/products/detectors. Accessed 27 May 2015
66. Dyakonov, M.I., Shur, M.S.: Shallow water analogy for a ballistic field effect transistor: new mechanism of plasma wave generation by dc current. Phys. Rev. Lett. **71**, 2465 (1993)
67. Dyakonov, M., Shur, M.S.: Detection, mixing, and frequency multiplication of terahertz radiation by two-dimensional electronic fluid. IEEE Trans. Electron Devices **43**(3), 380–387 (1996)
68. Sun, Y.F., Sun, J.D., Zhou, Y., Tan, R.B., Zeng, C.H., Xue, W., Qin, H., Zhang, B.S., Wu, D.M.: Room temperature GaN/AlGaN self-mixing terahertz detector enhanced by resonant antennas. Appl. Phys. Lett. **98**, 252103 (2011)
69. Sun, J.D., Qin, H., Lewis, R.A., Sun, Y.F., Zhang, X.Y., Cai, Y., Wu, D.M., Zhang, B.S.: Probing and modelling the localized self-mixing in a GaN/AlGaN field-effect terahertz detector. Appl. Phys. Lett. **100**, 173513 (2012)
70. Sun, J.D., Sun, Y.F., Zhou, Y., Zhang, Z.P., Lin, W.K., C.H., Zeng, Wu, D.M., Zhang, B.S., Qin, H., Li, L.L., Xu, W.: Enhancement of terahertz coupling efficiency by improved antenna design in GaN/AlGaN HEMT detectors. AIP Conf. Proc. **1399**, 893 (2011)
71. Zhou, Y., Sun, J.D., Sun, Y.F., Zhang, Z.P., Lin, W.K., Lou, H.X., Zeng, C.H., Lu, M., Cai, Y., Wu, D.M., Lou, S.T., Qin, H., Zhang, B.S.: Characterization of a room temperature terahertz detector based on a GaN/AlGaN HEMT. J. Semicond. **32**(4), 064005 (2011)
72. Sun, J.D., Qin, H., Lewis, R.A., Yang, X.X., Sun, Y.F., Zhang, Z.P., Li, X.X., Zhang, X.Y., Cai, Y., Wu, D.M., Zhang, B.S.: The effect of symmetry on resonant and nonresonant photoresponses in a field-effect terahertz detector. Appl. Phys. Lett. **106**, 031119 (2015)

Chapter 2
Field-Effect Self-Mixing Mechanism and Detector Model

Abstract Self-mixing of terahertz electromagnetic wave occurs in a field-effect electron channel when the terahertz electric field modulates both the local electron density and the drift velocity. In order to realize sensitive terahertz detection, asymmetry in the electric field and/or the charge density is required for generation of a unidirectional photocurrent/voltage. Existing hydrodynamic detection theories are reviewed and discussed. A detector model taking into account the spatial distributions of both the terahertz electric field and the electron density in the gated electron channel is developed in this chapter. The model presents full detector characteristics when both a source–drain bias and a gate voltage are applied. The model suggests that an asymmetric distribution of terahertz electric field is preferred for high-responsivity terahertz detection without a source–drain bias. The strength of terahertz photoresponse is characterized by the self-mixing factor and the field-effect factor. The former factor can be optimized by a strongly asymmetric and enhanced terahertz near field by using asymmetric terahertz antennas. Simulations based on the FDTD method confirm the effectiveness of asymmetric antenna design and the low-pass filter to isolate the antenna blocks from the electrical bonding pads for the detector.

2.1 The Physics of Terahertz Plasmon Detection

Plasmon in low-dimensional semiconductor structures has long been considered as an active medium for the detection and emission of terahertz electromagnetic waves. Plasmons as collective charge-density waves can be manipulated by tuning the electron density and by constructing specific plasma-wave cavities. In gated electron channels based on high-electron-mobility, two-dimensional electron gas (2DEG) tunable plasmon cavities can be formed. The interactions among the plasma wave, the terahertz electromagnetic wave, and the electron transport may be implemented in devices for high-sensitivity terahertz detection and high-efficiency terahertz emission. In this section, a review of the existing theories on plasmons in field-effect electron channel for detection is presented.

M.I. Dyakonov and M.S. Shur proposed terahertz emission in 1993 [1] and detection in 1996 [2] in gated 2DEG devices. The field-effect channel with a gate length

J. Sun, *Field-effect Self-mixing Terahertz Detectors*, Springer Theses,
DOI 10.1007/978-3-662-48681-8_2

of L defines the plasmon cavity. Assuming the plasmon cavity has a 'short-circuit' boundary on the left side and an 'open-circuit' boundary on the right side, the frequencies of the eigen plasmon modes are given as

$$\omega_N = (2N-1)\omega_0, \tag{2.1}$$

$$\omega_0 = \frac{\pi}{2L}\sqrt{\frac{e(U_\mathrm{g}-U_\mathrm{th})}{m^*}}, \tag{2.2}$$

where $N = 1, 2, \ldots$. Each eigen mode corresponds to an odd number of quarter wave length, i.e.,

$$L = (2N-1)\frac{\lambda_\mathrm{p}}{4}, \tag{2.3}$$

$$\lambda_\mathrm{p} = \frac{2\pi s}{\omega_\mathrm{p}}, \tag{2.4}$$

$$s = \sqrt{\frac{e(U_\mathrm{g}-U_\mathrm{th})}{m^*}}, \tag{2.5}$$

where λ_p is the plasma wave length, s is the plasma wave velocity, U_g is the gate voltage, U_th is the threshold voltage to pinch-off the 2DEG, e and m^* are the electron charge and effective mass, respectively. The plasmon modes can be tuned by the size of the cavity, i.e., the gate length L, and the electron density through the gate voltage. In a submicrometer-sized plasmon cavity, the plasmon frequency can be tuned into terahertz frequency range [2].

The asymmetric boundary condition and hence the quarter-wavelength nature of the plasmon cavity are preferred for both terahertz emission and detection. For emitters, self-excitation of plasma oscillations may be induced by electrons moving through the plasmon cavity under an external DC electric field [1]. For detectors, the asymmetry boundary conditions are essential to the generation of a unidirectional photocurrent or photovoltage. An intuitive picture of plasmon detection can be illustrated as follows. A terahertz electromagnetic wave with frequency ω excites resonantly/nonresonantly plasma waves in the channel. Electrons in the channel driven by the longitudinal electric field accompanied with the plasma wave under asymmetric boundary conditions induce a unidirectional current through the channel or equivalently a DC voltage across the source and the drain electrodes [2].

The excitation of plasmon modes in the gated electron channel depends on the quality factor of the specific mode

$$Q_\mathrm{p} = \omega_\mathrm{p}\tau_\mathrm{p}, \tag{2.6}$$

where, τ_p is the lifetime of the plasmon mode. The higher the quality factor, the easier the plasmon can be excited by either the incident terahertz wave or drifting electrons through the cavity. Resonant excitation of plasmon can be achieved when $Q_\mathrm{p} = \omega_\mathrm{p}\tau_\mathrm{p} \gg 1$. In contrast, plasmon modes are overdamped when $Q_\mathrm{p} = \omega_\mathrm{p}\tau_\mathrm{p} \ll 1$.

For the existing semiconductors, the quality factor is usually at the intermediate level, i.e., $\omega_p \tau_p \approx 1$–10. The nonresonant terahertz detection was observed in a GaAs HEMT device for the first time in 1998 [3]. Later on, nonresonant detection was achieved in short channel HEMT devices and CMOS devices [4–8]. As shown in Fig. 2.1, a typical nonresonant response is tuned by the gate voltage at 300 K [7]. Meanwhile, resonant terahertz detection was demonstrated in GaAs HEMTs and double-quantum-well (DQW) FETs [9–14]. A typical resonant response at 10 K as a function of the gate voltage is shown in Fig. 2.2 [10].

The physics behind the observed terahertz photoresponse can be illustrated in a model device where the terahertz field signal is applied between the gate and the source, as shown in Fig. 2.3. The nonlinear electron transport in the field-effect channel altered by the terahertz electric field gives rise to a photovoltage/current across/through the channel. M.I. Dyakonov and M.S. Shur studied the detection of terahertz electromagnetic wave in a field-effect transistor without an externally applied source–drain bias, i.e., in the linear regime of the field-effect transistor [2]. Veksler et al. extended the theory to the case when the electrons are driven by a DC current (a finite source–drain bias is applied between the drain and the source) [15]. It is found that the terahertz response can be strongly enhanced by the driving current, especially when the current reaches the saturation current. Here, we review the main results of these theoretical works as a starting point for our modeling on the terahertz detection in field-effect electron channel.

The governing equations for electrons in the channel are the hydrodynamic Euler equation of inertial motion and the continuity equation of electron flow

$$\frac{\partial v}{\partial t} + v\frac{\partial v}{\partial x} + \frac{v}{\tau} + \frac{e}{m^*}\frac{\partial U}{\partial x} = 0, \tag{2.7}$$

$$\frac{\partial n}{\partial t} + \frac{\partial (nv)}{\partial x} = 0, \tag{2.8}$$

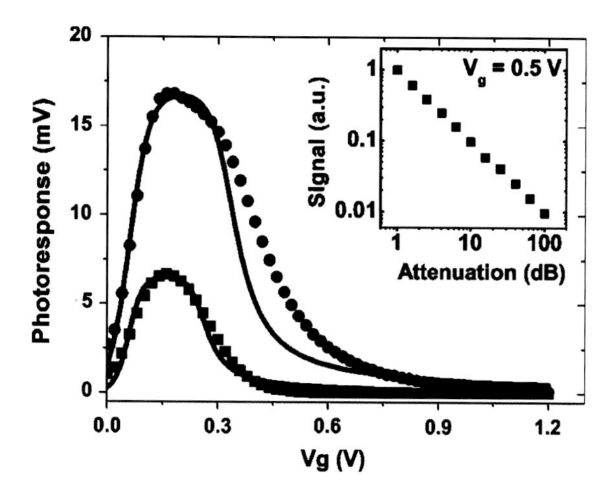

Fig. 2.1 Nonresonant terahertz response as a function of the gate voltage for two transistors with 600 nm (*squares*) and 800 nm (*circles*) gates at 120 GHz. *Inset* shows the signal versus attenuation for the gate voltage 0.5 V. Reprinted with permission of Ref. [7], copyright 2004, American Institute of Physics

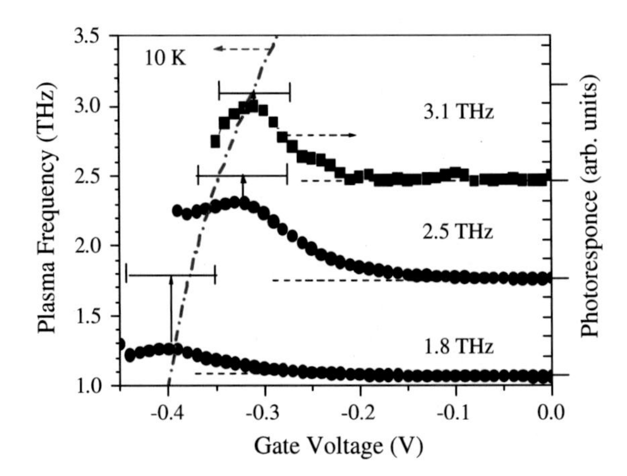

Fig. 2.2 Resonant terahertz response of an InGaAs device as a function of the gatevoltage at three different terahertz frequencies (1.8, 2.5 and 3.1 THz) at 10 K (*right axis*). Curves are shifted vertically for clarity. *Dashed lines* indicate the zero of the photoresponse. *Arrows* indicate resonance positions. Calculated plasmon frequency as a function of the gate voltage for $U_{\text{th}} = -0.41$ V is shown as the *dash-dotted line* (*left axis*). The *error bars* correspond to the linewidth of the observed resonance peaks. Reprinted with permission of Ref. [10], copyright 2006, American Institute of Physics

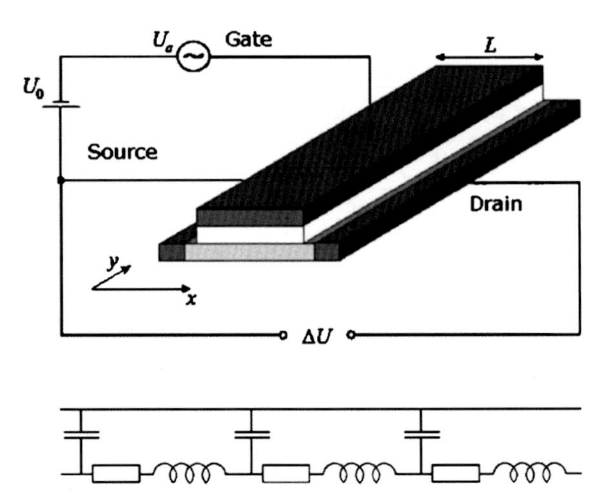

Fig. 2.3 A schematic and an equivalent circuit for the detector. Reprinted from Ref. [6], with kind permission of Springer Science+Business Media

where U is the local electrical potential in the channel, $-\partial U/\partial x$ is the local longitudinal electric field in the channel, v is the local electron velocity in direction of x, n is the local electron velocity, τ is the momentum relaxation time, e is the absolute value of the electron charge and m^* is the effective electron mass. In gradual channel approximation (GCA), the local electron density n in the channel is related to the local channel potential U

$$n = \frac{C_g(U_g - U_{th} - U)}{e}, \quad (2.9)$$

where, C_g is the gate to channel capacitance per unit area, U_g is the gate–source voltage and U_{th} is the threshold gate voltage to pinch off the electron channel. Equation 2.8 can be written as

$$\frac{\partial U}{\partial t} + \frac{\partial (Uv)}{\partial x} = 0. \quad (2.10)$$

When a terahertz signal with a frequency of ω is applied between the source and the gate, plasma wave can be excited at the source side and travels toward the drain side with a speed of $s = \sqrt{e(U_g - U_{th})/m^*}$. The boundary conditions for Eqs. (2.7) and (2.10) can be different depending on whether a finite source–drain bias is applied or not. In the case when no source–drain bias is applied and the drain is connected to ground, the boundary conditions can be rewritten as

$$U(0) = U_g - U_{th} + U_a \cos \omega t, \quad (2.11)$$

$$v(L) = 0. \quad (2.12)$$

where U_a is the potential induced by the incident terahertz wave. In the case when a source–drain bias is applied and a source–drain current I_d flows through the electron channel, the boundary conditions can be written as [1, 2]

$$U(0) = U_g - U_{th} + U_a \cos \omega t, \quad (2.13)$$

$$v(L) = I_d/U(L)C_g W, \quad (2.14)$$

where W is the channel width. The above boundary conditions mean that the terahertz signal is fed into the detector only from the source side. This asymmetric condition can be realized by using various methods, e.g., asymmetric source and drain contacts, different source–gate and drain–gate distances, and asymmetric terahertz antennas as will be introduced in Sect. 2.3.2.

Equations (2.7) and (2.10) with the corresponding boundary conditions can be solved either analytically or numerically and the induced photovoltage between the source and drain can be obtained. It is not necessary to repeat the results for all situations. Here, we introduce the results for the case when no external source–drain bias is applied. For resonant situation in which $\omega\tau \gg 1$ and $s\tau/L \gg 1$, i.e., the damping of the excited plasma wave is small and the gated channel forms a cavity to support discrete plasma wave modes according to Eq. (2.1), the photovoltage as a function of the terahertz frequency has a Lorentzian line shape

$$\delta U \propto \left(\frac{s\tau}{L}\right)^2 \frac{1}{1 + 4(\omega - \omega_N)^2\tau^2}. \quad (2.15)$$

In contrast, when $\omega\tau \ll 1$, i.e., the plasma wave is overdamped and the resonance is suppressed, the induced photovoltage is rather a smooth function of the frequency and the channel length

$$\delta U = \frac{1}{4}\frac{U_\mathrm{a}^2}{U_\mathrm{g} - U_\mathrm{th}}\frac{\sinh^2\kappa - \sin^2\kappa}{\sinh^2\kappa + \cos^2\kappa}, \tag{2.16}$$

where, $\kappa = (L/s)(\omega/2\tau)^2$. In the special case when the channel length is much larger than the decay length of the plasma wave

$$L \gg s(\frac{\tau}{\omega})^2, \tag{2.17}$$

the induced photovoltage has the form

$$\delta U = \frac{1}{4}\frac{U_\mathrm{a}^2}{U_\mathrm{g} - U_\mathrm{th}}\left\{1 + \frac{2\omega\tau}{\sqrt{1+\omega^2\tau^2}}\right\}, \tag{2.18}$$

i.e., independent on the channel length.

The photovoltage strongly depends on the nonlinearity in the field-effect controlled electron density. In the case when a source–drain bias is applied and a DC current $I_\mathrm{d} \neq 0$ flows through the channel, the electron density varies from the source side to the drain side. Such a nonuniform spatial distribution of the electron density enhances the photoresponse. According to the theoretical analysis done by Veksler et al., the photovoltage tends to be infinitely high when the source–drain current reaches the saturation current [15]. For example, the $I - V$ characteristic and the nonresonant response at 200 GHz of a GaAs HEMT are shown in Fig. 2.4a and b,

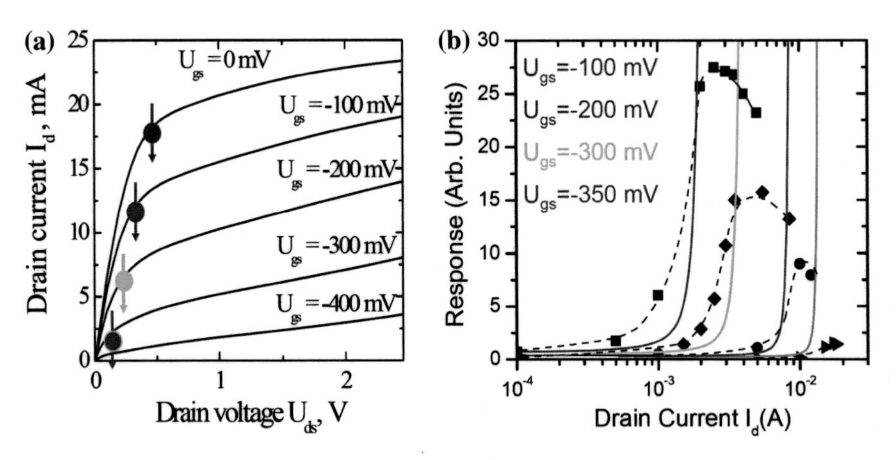

Fig. 2.4 **a** $I - V$ characteristics of a GaAs HEMT at different gate voltages. The *dots* correspond to the saturation voltage. **b** Terahertz response at 200 GHz as a function of the drain current. *Solid curves* represent the theoretical photovoltage. Reprinted with permission of Ref. [15], copyright 2006, American Physical Society

respectively. The photovoltage increases significantly, then reaches a maximum and finally decreases with increasing the drain current. Such a behavior is not recovered by the calculated photoresponse plotted as the solid curves in Fig. 2.4b.

The above theories provide clear pictures for different terahertz detection processes in the linear and saturation regimes of a field-effect transistor. However, they are inefficient in providing quantitative analyses of the photovoltage. The assumption that the terahertz signal is applied between the gate and the source and/or between the drain and the source neglects the detailed spatial distribution of the terahertz field. Possible interplay between the terahertz field and the electron channel with a nonuniform density is neglected as well. Therefore, it is of great importance to develop a detector model that takes into account both the spatial distributions of the terahertz field and the electron density. Such a detector model would provide full description of the photoresponses in the linear regime (LR), the saturation regime (SR), and the transition regime (TR). Above all, the model would provide guidance for the design and optimization of field-effect terahertz detectors.

2.2 Quasi-Static Self-Mixing Detector Model

As discussed in the previous section, the theories of terahertz detection based on field-effect transistors (FETs) assume an asymmetric terahertz field distribution across the gated electron channel in the first place. A clear physical picture of self-mixing mechanism is given by the resonant and nonresonant theories. However, no design rule for detector optimization is given. Furthermore, the theories do not clearly provide detector characteristics when both a gate voltage and a source–drain bias are applied [15]. Here, we present a detector model taking into account both the spatial distributions of the terahertz field and the charge density in the gated channel. The model is of quasi-static and is based on the gradual channel approximation (GCA). This model provides a full description of the detector response in the linear regime (LR), the saturation regime (SR), and the transition regime (TR). Optimized detector design can be made based on this model [16–22].

A schematic of the model detector and the equivalent circuit are given in Fig. 2.5. The active region of the detector is composed of a field-effect electron channel and terahertz antennas. According to GCA, the field-effect electron channel can be divided into a series of infinitely small segments of the electron channel. Under a source–drain bias V_{ds}, the current through the channel can be expressed as [23]

$$i_x = -eWn_x\mu\frac{\mathrm{d}V_x}{\mathrm{d}x} = -eWn_x(V_{\mathrm{geff}})\mu\frac{\mathrm{d}V_x}{\mathrm{d}x}, \tag{2.19}$$

where μ is the electron mobility, W is the channel (2DEG mesa) width. We define $V_{\mathrm{geff}} = V_{\mathrm{g}} - V_{\mathrm{th}} - V_x$ as the effective gate voltage at location x, where V_{g} and V_{th} are the applied DC gate voltage and the threshold gate voltage to pinch off the 2DEG. The channel potential V_x varies from 0 to V_{ds} from $x = 0$ to $x = L$. The local electron

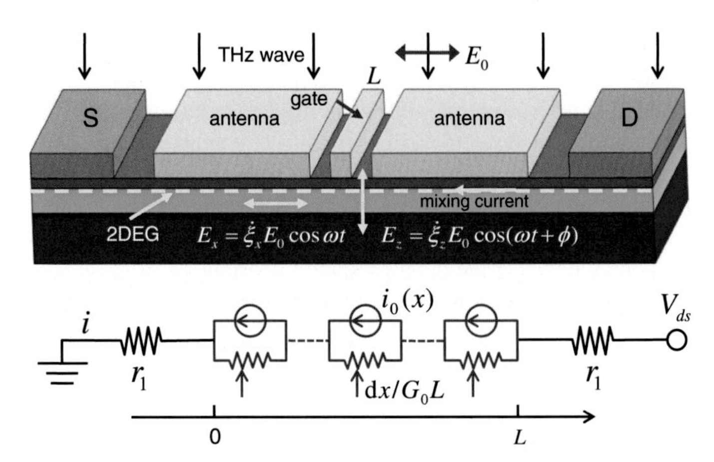

Fig. 2.5 A schematic and an equivalent circuit of the self-mixing detector. Reprinted with permission of Ref. [17], copyright 2012, American Institute of Physics

density is controlled by the effective local gate voltage: $n_x = C_g V_{geff}/e$, where C_g is the effective gate-channel capacitance per unit area. The total current through the electron channel can be expressed as

$$I_{ds} = \frac{1}{L}\int_0^L i_x \, dx = -\frac{eW\mu C_g}{L}\int_0^{V_{ds}} V_{geff} \, dV_x. \tag{2.20}$$

The drain–source current and the local channel potential can be further written as

$$I_{ds} = \begin{cases} \mu C_g W\left[2(V_g - V_{th})V_{ds} - V_{ds}^2\right]/2L & \text{if } V_{ds} \le V_g - V_{th}, \\ \mu C_g W\left[(V_g - V_{th})^2 + \lambda(V_{ds} - V_g + V_{th})\right]/2L & \text{if } V_{ds} > V_g - V_{th}, \end{cases} \tag{2.21}$$

$$V_x = \begin{cases} (V_g - V_{th})\left[1 - (1 - x/L_{LR})^{1/2}\right], x = [0, L] & \text{if } V_{ds} \le V_g - V_{th}, \\ (V_g - V_{th})\left[1 - (1 - x/L_{SR})^{1/2}\right], x = [0, L_{SR}] & \text{if } V_{ds} > V_g - V_{th}, \end{cases} \tag{2.22}$$

where, parameter λ describes the degree of the effective channel-length modulation $L \to L_{SR} = L/[1 + \lambda(V_{ds} - V_g + V_{th})/(V_g - V_{th})^2]$ in regime SR, and $L_{LR} = L(V_g - V_{th})^2/[2(V_g - V_{th})V_{ds} - V_{ds}^2]$ in regime LR.

From Eqs. (2.19), (2.21) and (2.22), the charge density and its derivative dn_x/dV_{geff} defined as field-effect factor at any location can be numerically calculated. In the case when $V_{ds} = 0\,V$, the differential conductance can be simply written as

$$G_0 = e\mu\frac{W}{L}n(V_g - V_{th}). \tag{2.23}$$

For a realistic field-effect detector, there exists resistances from the ohmic contacts and the 2DEG mesa in series with the gated electron channel. These series resistances are absorbed into $2r_1$, as shown in Fig. 2.5.

Upon terahertz irradiation with a frequency of $\omega = 2\pi f$ and an energy flux of P_0, both a horizontal ($\dot{\xi}_x E_0$) and a perpendicular ($\dot{\xi}_z E_0$) terahertz field in the channel are induced. Under the quasi-static assumption, i.e., $\omega\tau \ll 1$, the local effective channel potential and the effective gate voltage are modulated by the terahertz field and can be written as

$$V_x \rightarrow V_x + \xi_x E_0 \cos(\omega t), \tag{2.24}$$

$$V_\mathrm{g} \rightarrow V_\mathrm{g} + \xi_z E_0 \cos(\omega t + \phi), \tag{2.25}$$

where $\dot{\xi}_x = \mathrm{d}\xi_x/\mathrm{d}x$, $\dot{\xi}_z = \mathrm{d}\xi_z/\mathrm{d}z$ and ϕ are the horizontal and perpendicular field enhancement factors, and the phase difference between the induced fields, respectively. E_0 is determined by the incident energy flux and the free-space impedance Z_0 by $P_0 = E_0^2/2Z_0$.

Substituting Eqs. (2.24) and (2.25) into Eq. (2.19), we can get the current through the electron channel

$$i_x = e\mu W n(V_\mathrm{geff} + \xi_z E_0 \cos(\omega t + \phi) - \xi_x E_0 \cos\omega t)\frac{\mathrm{d}}{\mathrm{d}x}(V_x + \xi_x E_0 \cos\omega t). \tag{2.26}$$

We assume that the total current can be obtained as

$$i_\mathrm{ds} = \frac{1}{LT}\int_0^T \mathrm{d}t \int_0^L \mathrm{d}x\, i_x. \tag{2.27}$$

Where $T = 2\pi/\omega$ is the terahertz oscillating period. This assumption is valid for the case when the channel length is larger than the decay length of the plasma wave or in general the plasma wave excitation is confined in a small fraction of the gated channel. Furthermore, it is safely assumed that the incident terahertz power is low and the induced terahertz potential is small comparing to the applied gate voltage and the source-drain bias. The electron density can be expressed as

$$n = n(V_\mathrm{geff}) + \frac{\mathrm{d}n}{\mathrm{d}V_\mathrm{geff}}\xi_z E_0 \cos(\omega t + \phi) - \frac{\mathrm{d}n}{\mathrm{d}V_\mathrm{geff}}\xi_x E_0 \cos\omega t. \tag{2.28}$$

By integrating i_x along the whole gated electron channel, we have

$$\begin{aligned} &\frac{1}{L}\int_0^L i_x\, \mathrm{d}x \\ &= \frac{e\mu W}{L}\int_0^L \left[n(V_\mathrm{geff}) + \frac{\mathrm{d}n}{\mathrm{d}V_\mathrm{geff}}\xi_z E_0 \cos(\omega t + \phi) - \frac{\mathrm{d}n}{\mathrm{d}V_\mathrm{geff}}\xi_x E_0 \cos\omega t\right] \\ &\quad \times \left[\mathrm{d}V_x + \dot{\xi}_x E_0 \cos\omega t\, \mathrm{d}x\right]. \end{aligned} \tag{2.29}$$

The time-averaged source-drain current can be obtained by averaging Eq. (2.29) in the oscillation period

$$i_{\mathrm{ds}} = I_{\mathrm{ds}} + i_{xz} + i_{xx}, \tag{2.30}$$

where, the first term is the DC bias current the same as that in Eq. (2.21), the rest are two different types of self-mixing current. The self-mixing photocurrent can be expressed as

$$i_{xz} = \frac{e\mu W}{2L} Z_0 P_0 \bar{z} \int_0^L \frac{\mathrm{d}n}{\mathrm{d}V_{\mathrm{geff}}} \dot{\xi}_x \dot{\xi}_z \cos\phi \, \mathrm{d}x, \tag{2.31}$$

$$i_{xx} = -\frac{e\mu W}{2L} Z_0 P_0 \int_0^L \frac{\mathrm{d}n}{\mathrm{d}V_{\mathrm{geff}}} \xi_x \dot{\xi}_x \, \mathrm{d}x, \tag{2.32}$$

where $\bar{z} = \xi_z / \dot{\xi}_z$ is the effective distance between the gate and the 2DEG. Current $i_{xz} \propto \dot{\xi}_x \dot{\xi}_z$ is induced by both the horizontal and the perpendicular fields, while $i_{xx} \propto \xi_x \dot{\xi}_x$ is only from the horizontal field which is similar to that in a SBD.

The responsivity can be greatly enhanced by optimizing the channel electron mobility, channel geometry, and antenna efficiency. The current responsivity can be expressed as

$$R_{\mathrm{i}} = \frac{i_{xz} + i_{xx}}{P_0} = \frac{e\mu W}{2L} Z_0 \left\{ \bar{z} \int_0^L \frac{\mathrm{d}n}{\mathrm{d}V_{\mathrm{geff}}} \dot{\xi}_x \dot{\xi}_z \cos\phi \, \mathrm{d}x - \int_0^L \frac{\mathrm{d}n}{\mathrm{d}V_{\mathrm{geff}}} \xi_x \dot{\xi}_x \, \mathrm{d}x \right\}. \tag{2.33}$$

Equations (2.31) and (2.32) can be numerically calculated as a function of the gate voltage and the source-drain bias. As will be shown in Chap. 3, i_{xz} is usually the dominant self-mixing signal. For the special case when $V_{\mathrm{ds}} = 0\,\mathrm{V}$, according to Eq. (2.23), the effective gate voltage and the electron density are $V_{\mathrm{geff}} = V_{\mathrm{g}} - V_{\mathrm{th}}$ and $n = G_0 L / e\mu W$, respectively. The total self-mixing current under zero source–drain bias can be written as

$$i_0 = \frac{Z_0 P_0}{2} \frac{\mathrm{d}G_0}{\mathrm{d}V_{\mathrm{geff}}} \left\{ \bar{z} \int_0^L \dot{\xi}_x \dot{\xi}_z \cos\phi \, \mathrm{d}x - \int_0^L \xi_x \dot{\xi}_x \, \mathrm{d}x \right\}. \tag{2.34}$$

The integral is independent on the gate voltage and is determined by the terahertz electric field distribution. Antenna is commonly used to manipulate the terahertz electric field. Hence, we define the integral as antenna factor A_0 which characterizes the enhancement of the terahertz electric fields,

$$A_0 = \bar{z} \int_0^L \dot{\xi}_x \dot{\xi}_z \cos\phi \, \mathrm{d}x - \int_0^L \xi_x \dot{\xi}_x \, \mathrm{d}x. \tag{2.35}$$

The self-mixing photocurrent under zero bias can be simplified as

$$i_0 = A_0 \times \frac{Z_0 P_0}{2} \frac{\mathrm{d}G_0}{\mathrm{d}V_{\mathrm{geff}}}. \tag{2.36}$$

According to the Ohm's law, the photovoltage under zero bias can be expressed as

$$v_0 = A_0 \times \frac{Z_0 P_0}{2G_0} \frac{\mathrm{d}G_0}{\mathrm{d}V_{\mathrm{geff}}}. \tag{2.37}$$

Similarly, under zero source–drain bias, the responsivity is in a simple form

$$R_{\mathrm{i}} = A_0 \times \frac{Z_0}{2} \frac{\mathrm{d}G_0}{\mathrm{d}V_{\mathrm{geff}}}. \tag{2.38}$$

As shown in Fig. 2.6, the effects of the series resistance of the detector and the input impedance of the voltmeter (r_{mv}) or the current meter (r_{mc}) affect the measurement of the photocurrent or the photovoltage. The measured conductance including the series resistance is $G_{\mathrm{m}} = G_0/(1 + 2r_1 G_0)$. The current meter and the voltmeter measure only a fraction of the photocurrent i_0 and the photovoltage $v_0 = i_0/G_0$. The photocurrent and photovoltage under zero bias can be expressed as

$$i_{\mathrm{m}} = A_0 \times \frac{Z_0 P_0}{2} \frac{1}{(2r_1 + r_{\mathrm{mc}})(1 - 2r_1 G_{\mathrm{m}})G_{\mathrm{m}} + (1 - 2r_1 G_{\mathrm{m}})^2} \frac{\mathrm{d}G_{\mathrm{m}}}{\mathrm{d}V_{\mathrm{geff}}}, \tag{2.39}$$

$$v_{\mathrm{m}} = A_0 \times \frac{Z_0 P_0}{2} \frac{r_{\mathrm{mv}}}{(2r_1 + r_{\mathrm{mv}})(1 - 2r_1 G_{\mathrm{m}})G_{\mathrm{m}} + (1 - 2r_1 G_{\mathrm{m}})^2} \frac{\mathrm{d}G_{\mathrm{m}}}{\mathrm{d}V_{\mathrm{geff}}}. \tag{2.40}$$

When the detector noise is dominated by the thermal noise, the noise-equivalent power (*NEP*) of the detector can be expressed as

$$NEP = \begin{cases} \frac{N_{\mathrm{iB}}}{R_{\mathrm{i}}} = \frac{\sqrt{4k_{\mathrm{B}}TG_{\mathrm{m}}}}{i_{\mathrm{m}}/P_0}, & \text{for photocurrent readout,} \\ \frac{N_{\mathrm{vB}}}{R_{\mathrm{v}}} = \frac{\sqrt{4k_{\mathrm{B}}T/G_{\mathrm{m}}}}{v_{\mathrm{m}}/P_0}, & \text{for photovoltage readout,} \end{cases}$$

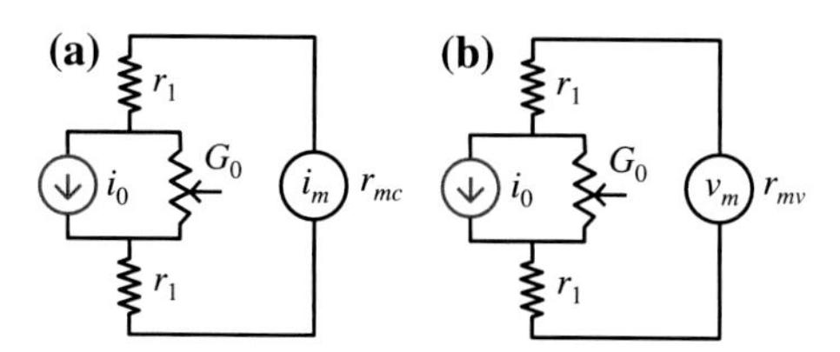

Fig. 2.6 Equivalent circuits for the detector when its **a** short-circuit photocurrent and **b** open-circuit photovoltage is considered, respectively. The series resistance of the detector (r_1) and the input impedance of the amplifiers (r_{mc} and r_{mv}) are included as well

where, T is the temperature, k_B is the Boltzmann constant, N_{iB} and N_{vB} are spectral noise current and noise voltage, respectively. The measured current responsivity and the measured voltage responsivity are defined as $R_i = i_m/P_0$ and $R_v = v_m/P_0$, respectively. It is clear that the series resistance decreases the responsivity and increases the noise-equivalent power.

2.3 Simulation of Terahertz Antenna by Using FDTD Method

Based on the above quasi-static self-mixing model, the responsivity is determined by two factors. One of the factors is directly related to the degree of controlling the electron density by the gate voltage $\mathrm{d}n/\mathrm{d}V_{\mathrm{geff}}$, which is all about the field-effect and can be improved by enhancing the electron mobility, the gate capacitance and the geometry of the gate. The other factor is named as self-mixing factor $\dot{\xi}_x\dot{\xi}_z \cos\phi$, which is determined by the design of the antennas. Therefore, an optimized design of the antennas becomes crucial to increase the responsivity. In this section, the Finite-Difference in Time Domain (FDTD) method is introduced for antenna simulations.

2.3.1 *Principle of FDTD and the Algorithm*

In general, FDTD methods belong to the class of finite-element methods for numerically solving differential equations [24]. Proposed by Yee in 1966 [25], FDTD has been developed rapidly into a wide range of tools and applications such as FDTD Solutions, CST Microwave Studio, HFSS, EastFDTD, etc. FDTD solutions cover a wide frequency range with a single simulation run and treat nonlinear material properties in a natural way. FDTD method is widely used for numerical analysis of electrodynamics.

The time-dependent Maxwell's equations are discretized using central-difference approximations to the space and time partial derivatives. The resulting finite-difference equations are solved in either software or hardware in a leapfrog manner. The electric field vector components in a volume of space are solved at a given instant of time. Then the magnetic field vector components in the same spatial volume are solved at the next instant of time. These two processes are repeated over and over until the desired transient or steady-state electromagnetic field behavior is fully evolved.

It is worthwhile to introduce the method of finite difference. Setting the variable $\Delta x = h$ for function $f(x)$, the forward difference of $f(x)$ is

$$\Delta f = f(x+h) - f(x). \tag{2.41}$$

The derivative of $f(x)$ can be expressed in terms of the forward difference, the backward difference, and the center difference

$$\frac{\mathrm{d}f}{\mathrm{d}x} = \lim_{\Delta x \to 0} \frac{\Delta f(x)}{\Delta x} \simeq \begin{cases} \frac{f(x+h)-f(x)}{h} & \text{forward} \\ \frac{f(x)-f(x-h)}{h} & \text{backward} \\ \frac{f(x+h)-f(x-h)}{2h} & \text{center} \end{cases} \tag{2.42}$$

According to the Taylor expansion, we get

$$f(x+h) = f(x) + h\frac{\mathrm{d}f(x)}{\mathrm{d}x} + \frac{1}{2!}h^2\frac{\mathrm{d}^2 f(x)}{\mathrm{d}x^2} + \frac{1}{3!}h^3\frac{\mathrm{d}^3 f(x)}{\mathrm{d}x^3} + \cdots \tag{2.43}$$

$$f(x-h) = f(x) - h\frac{\mathrm{d}f(x)}{\mathrm{d}x} + \frac{1}{2!}h^2\frac{\mathrm{d}^2 f(x)}{\mathrm{d}x^2} - \frac{1}{3!}h^3\frac{\mathrm{d}^3 f(x)}{\mathrm{d}x^3} + \cdots \tag{2.44}$$

$$f(x+h) - f(x-h) = 2h\frac{\mathrm{d}f(x)}{\mathrm{d}x} + \frac{2}{3!}h^3\frac{\mathrm{d}^3 f(x)}{\mathrm{d}x^3} + \cdots . \tag{2.45}$$

The central-difference approximation has an error which is proportional to the quadratic term of h.

The time-dependent Maxwell's equations are

$$\nabla \times \mathbf{H} = \frac{\partial \mathbf{D}}{\partial t} + \mathbf{J} \tag{2.46}$$

$$\nabla \times \mathbf{E} = -\frac{\partial \mathbf{B}}{\partial t} - \mathbf{J}_m \tag{2.47}$$

where $\mathbf{E}$ is the electric field, $\mathbf{D}$ is the displacement field, $\mathbf{H}$ is the magnetic field strength, $\mathbf{B}$ is the magnetic flux density, $\mathbf{J}$ is the current density, $\mathbf{J}_m$ is the magnetic flux density, ε is the dielectric constant, μ is the magnetic permeability, σ is the electrical conductivity, σ_m is the magnetic permeability.

In Cartesian coordinates, Eq. (2.47) can be written as

$$\begin{aligned} \frac{\partial H_z}{\partial y} - \frac{\partial H_y}{\partial z} &= \varepsilon\frac{\partial E_x}{\partial t} + \sigma E_x, \\ \frac{\partial H_x}{\partial z} - \frac{\partial H_z}{\partial x} &= \varepsilon\frac{\partial E_y}{\partial t} + \sigma E_y, \\ \frac{\partial H_y}{\partial x} - \frac{\partial H_x}{\partial y} &= \varepsilon\frac{\partial E_z}{\partial t} + \sigma E_z, \end{aligned} \tag{2.48}$$

and

$$\begin{aligned}\frac{\partial E_z}{\partial y} - \frac{\partial E_y}{\partial z} &= -\mu\frac{\partial H_x}{\partial t} - \sigma_m H_x,\\ \frac{\partial E_x}{\partial z} - \frac{\partial E_z}{\partial x} &= -\mu\frac{\partial H_y}{\partial t} - \sigma_m H_y,\\ \frac{\partial E_y}{\partial x} - \frac{\partial E_x}{\partial y} &= -\mu\frac{\partial H_z}{\partial t} - \sigma_m H_z.\end{aligned} \tag{2.49}$$

Equations (2.48) and (2.49) can be rewritten in a form of the central-difference. Define $f(x, y, z, t)$ as one component of **E** and **H** in Cartesian coordinates, and discretize it in time domain and the space domain with grid number n as

$$f(x, y, z, t) = f(i\Delta x, j\Delta y, k\Delta z, n\Delta t) = f^n(i, j, k). \tag{2.50}$$

The partial derivative can be approximated by the central differences,

$$\left.\frac{\partial f(x, y, z, t)}{\partial x}\right|_{x=i\Delta x} \simeq \frac{f^n(i+\frac{1}{2}, j, k) - f^n(i-\frac{1}{2}, j, k)}{\Delta x} \tag{2.51}$$

$$\left.\frac{\partial f(x, y, z, t)}{\partial y}\right|_{y=i\Delta y} \simeq \frac{f^n(i, j+\frac{1}{2}, k) - f^n(i, j-\frac{1}{2}, k)}{\Delta y} \tag{2.52}$$

$$\left.\frac{\partial f(x, y, z, t)}{\partial z}\right|_{z=i\Delta z} \simeq \frac{f^n(i, j, k+\frac{1}{2}) - f^n(i, j, k-\frac{1}{2})}{\Delta z} \tag{2.53}$$

Yee cellular in FDTD algorithm is shown in Fig. 2.7. Each magnetic field vector component is surrounded by four electric field vector components in a loop. Similarly, each electric field vector component is surrounded by four magnetic field vector components. This spatial sampling method of the electromagnetic components not only conforms to Faraday's law and Ampere's law, but is also suitable for difference calculation of the Maxwell equations and properly describes the propagation of

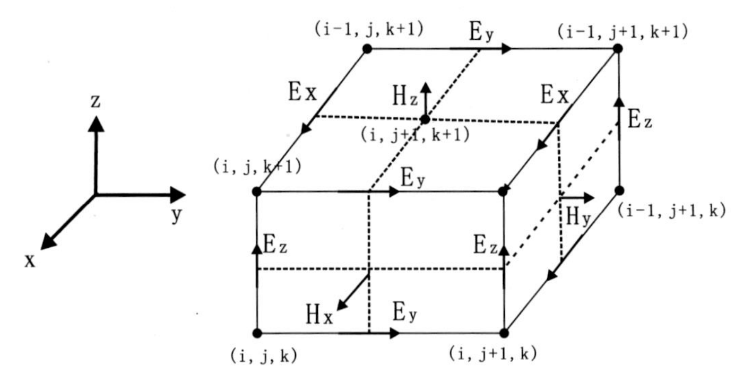

Fig. 2.7 Yee cellular in FDTD algorithm

electromagnetic wave. In addition, the sampling of the electric field and the magnetic field alternate in time domain, the sampling time intervals differ from each other by a half-time step. In this way, the Maxwell's equations become explicit difference equations and can be iteratively solved in the time domain. The space distribution of the electromagnetic field at each time step can be gradually calculated with given initial values and boundary conditions.

2.3.2 Antenna Simulation by FDTD

Dipole antennas have been widely used in microwave and radio frequency fields. Here, dipole antennas will be used to manipulate the near-field terahertz wave in the gated electron channel [16, 17]. FDTD simulations will be performed to evaluate the spatial distribution of the terahertz electromagnetic wave and its phase in the near-field zones of a symmetric and an asymmetric antenna.

A typical symmetric antenna on top of a GaN substrate is shown in Fig. 2.8. It contains a pair of metal dipole blocks each of which has a length of 45 μm and a width of 10 μm. The length is about a quarter of the wavelength of the terahertz electromagnetic wave at 900 GHz. The width of the electron channel (2DEG mesa) under the antenna is 4 μm. The field-effect gate with a length of 2 μm is inserted in between the two dipole blocks and the spacing between the gate and the antennas is 1.5 μm.

The spatial distributions of the terahertz electric field and the phase distributions at 900 GHz about 23 nm below the antennas are simulated using the FDTD method. The field enhancement factors for the horizontal and the perpendicular field are shown in Fig. 2.9a, b, respectively. The horizontal field is concentrated in the gaps between the gate and the antennas, while the perpendicular field is mainly distributed under the gate. The spatial distributions of the field strength are symmetric. However, the perpendicular field vanishes and changes its phase by π at $x \approx 0$ μm, while the horizontal field keeps its phase constant along the channel, as shown in Fig. 2.9c, d. In Fig. 2.9e, the simulated mixing factor is plotted to reveal the spatial distribution of self-mixing. Strong self-mixing with a factor of 2,000 occurs mainly in the edge areas of the gated electron channel. The length of the active area where self-mixing

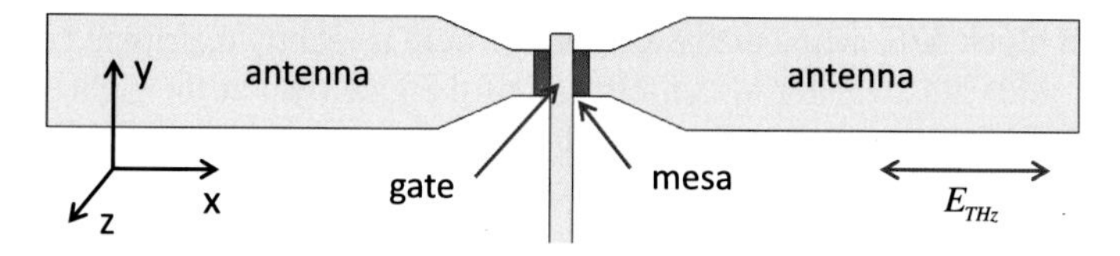

Fig. 2.8 A typical symmetric antenna with two dipole blocks and a gate. The 2DEG channel is under the gate and is connected to the dipole blocks

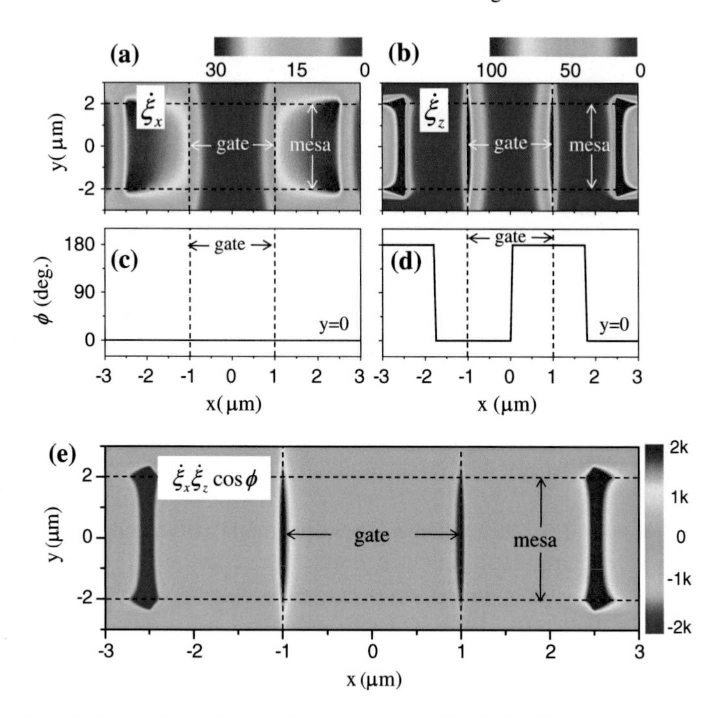

Fig. 2.9 Spatial distributions of the terahertz electric field and the phase simulated using FDTD method at 900 GHz. **a** Field enhancement factor $\dot{\xi}_x$ along the channel. **b** Field enhancement factor $\dot{\xi}_z$ perpendicular to the channel. **c**, **d** Phase ϕ_x and ϕ_z at $y = 0$. **e** Mixing factor $\dot{\xi}_x\dot{\xi}_z \cos\phi$

is predominant is about 200 nm and is only one-tenth of the gate length. Due to the phase change at $x = 0\,\mu\text{m}$, the mixing at the source side generates a photocurrent opposite to that induced at the drain side. With a symmetric design of the antennas, the overall photocurrent vanishes. Hence, a symmetric antenna design is not favorable for high sensitivity detection.

An asymmetric antenna design consisting of three dipole blocks is shown in Fig. 2.10. Similar to those shown in Fig. 2.8, each dipole block (A, B and C) is $45\,\mu\text{m} \times 10\,\mu\text{m}$, the gate length is $2\,\mu\text{m}$, and the gap between the gate and the two isolated blocks (A and B) is $1.5\,\mu\text{m}$. Block C is used to induce asymmetric terahertz electric field in the gated electron channel.

As shown in Fig. 2.11a, b, the horizontal field is concentrated in the gaps between the gate and block A/B, while the perpendicular field is mainly distributed under the gate. Both fields are stronger at the left side of the gate than at the right side. The horizontal field vanishes and changes its phase by π at $x \approx +0.3\,\mu\text{m}$, while the perpendicular field keeps its phase constant along the channel as shown in Fig. 2.11c, d. Similar to that shown in Fig. 2.9e, self-mixing is highly localized near the edges of the gated electron channel, as shown in Fig. 2.11e. The mixing factor $\dot{\xi}_x\dot{\xi}_z \cos\phi$ at the left side and the right side are about 10,000 and 2,000, respectively. Self-mixing is greatly enhanced by the asymmetric design. Most importantly, the asymmetric

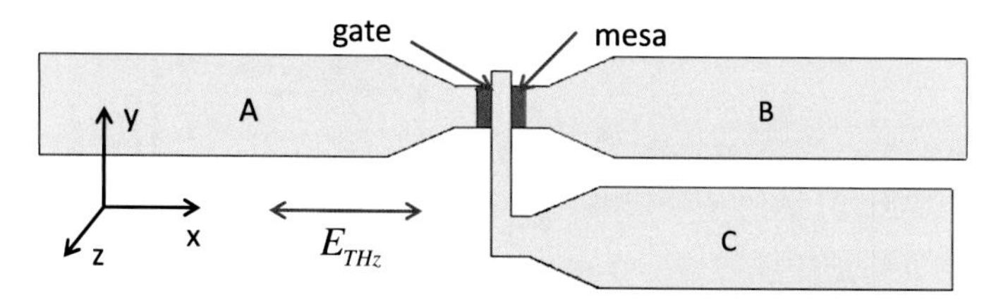

Fig. 2.10 Asymmetric antenna consists of three dipole blocks A, B, and C

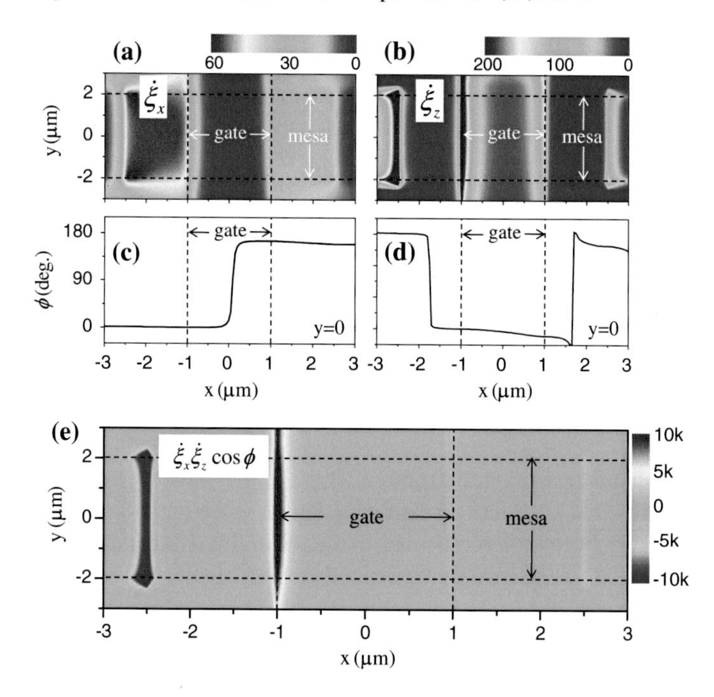

Fig. 2.11 Terahertz field distribution in the near-field zone of an asymmetric antenna designed for 900 GHz. Spatial distributions of the field enhancement factors (**a**, **b**) and the phases (**c**, **d**) of the horizontal and the perpendicular field. **e** Spatial distribution of the mixing factor $\dot{\xi}_x \dot{\xi}_z \cos\phi$

design allows for the generation of photocurrent predominantly determined by the self-mixing in one of the two active regions. Asymmetric antenna design is crucial for high responsivity terahertz detection.

The aspect ratio of the antenna width (AW) and the antenna length (AL) affects the self-mixing factor at the resonant frequency. The resonant frequency and the mixing factor as a function of the block length (AL) while the antenna width is fixed at 10 μm is shown in Fig. 2.12a. With increasing the antenna length from 10 μm to 120 μm, the mixing factor $\dot{\xi}_x \dot{\xi}_z \cos\phi$ is enhanced from 500 to 90,000 and the resonant frequency decreases from 1,740 to 380 GHz. The resonant frequency and the mixing factor as a function of the antenna width while the length width is fixed at 45 μm is shown in Fig. 2.12b. By increasing the antenna width from 4 to 20 μm, the mixing

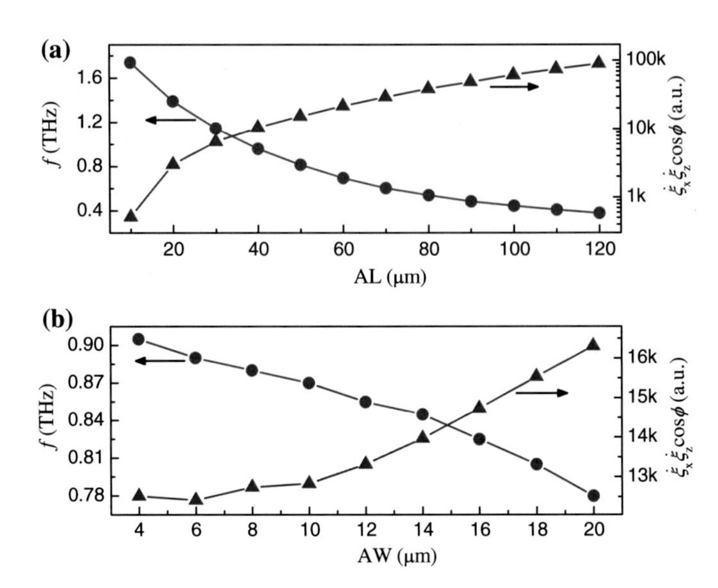

Fig. 2.12 Resonant frequency and mixing factor as a function of **a** the length (AL) and **b** the width (AW) of the antenna blocks. The width and the length are set as 10 and 45 μm for (**a**) and (**b**), respectively

factor $\dot{\xi}_x \dot{\xi}_z \cos\phi$ increases slightly from 12,500 to 16,300 and the resonant frequency decreases from 905 to 780 GHz. The antenna length is the main factor affecting the self-mixing factor. Furthermore, the mixing factor of this 3-block antenna designed for lower resonant frequencies is much larger than that at higher resonant frequencies.

2.3.3 Terahertz Filter

The resonant characteristics of the antenna are vulnerable to any nearby metallic structures, such as bonding pads and the interconnection between the pads and the antenna blocks. The effect of the bonding pads on the self-mixing factor is simulated and is shown in Fig. 2.13. The horizontal and the perpendicular electric field are about 400 and 20 at the resonant frequency 1 THz, respectively. In the case when the antennas and the bonding pads are connected via straight metal lines as shown in Fig. 2.13b, the maximum perpendicular field decreases from 400 to 180 and an extra resonance around 200 GHz is induced. To eliminate this negative effect of bonding pads on the mixing factor and hence the responsivity, we use low pass filters as interconnection between the antenna blocks and the bonding pads, as shown in Fig. 2.13c. Simulations indicate that by isolating the antennas from the bonding pads the self-mixing factor can be restored to its original value as that shown in Fig. 2.13a.

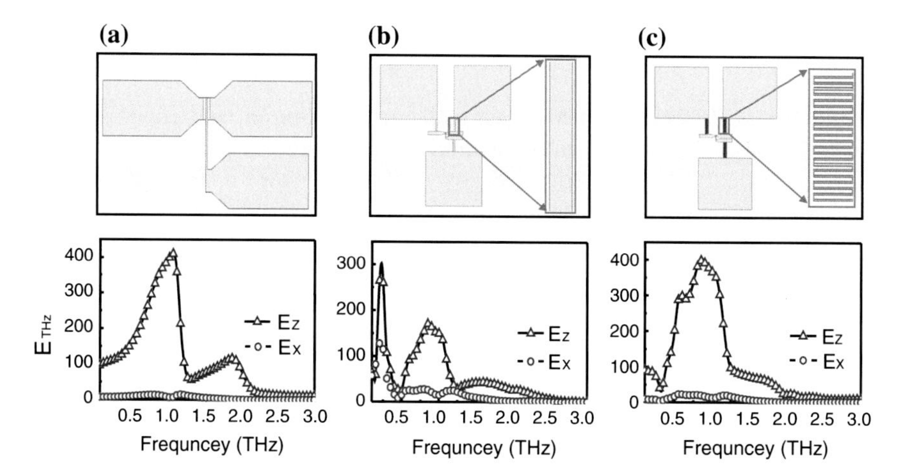

Fig. 2.13 Terahertz electric fields as a function of the frequency. **a** Asymmetric antenna without any bonding pad. **b** Asymmetric antenna with bonding pads connected to each antenna block by a straight metal line. **c** Asymmetric antenna with terahertz filters connecting the antenna blocks and the bonding pads. The filter is meander shaped. Reprinted with permission of Ref. [18], copyright 2011, American Institute of Physics

Therefore, the meander-shaped low pass filters connecting the antenna with the corresponding bonding pad can maintain the mixing factor and ensure a good performance for *NEP*. Simulations also suggest that the isolation effect can be enhanced by increasing the length of the filter. For the meander-shaped filter, the filter effect can be optimized when the periodicity is more than 25. A scanning electron microscope graph shown in Fig. 2.14 is such a terahertz filter with a line width of 150 nm, a width of 500 nm and a periodicity of 30.

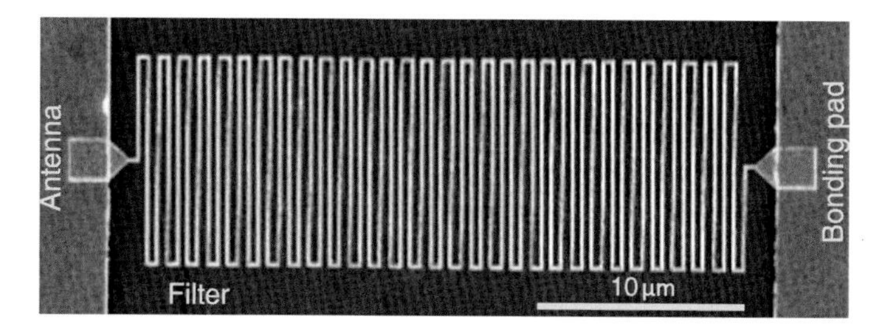

Fig. 2.14 Scanning electron microscope photograph of a meander shaped low pass filter

2.4 Summary

We summarize the existing theories on terahertz detection in field-effect electron channel based on the time-dependent hydrodynamic transport equation and the continuity equation with asymmetric boundary conditions either for the zero-biased or the biased situations. We develop a quasi-static self-mixing detector model based on the gradual channel approximation. The model takes into account the spatial distributions of terahertz electric fields and the electron density in the gated electron channel. Numerical calculations can be made based on this model to present quantitative terahertz photoresponse in the linear regime, the saturation regime, the transition regime, and the pinch-off regime. According to the model, the terahertz photoresponse is proportional to the local derivative of the electron density $\mathrm{d}n_x/\mathrm{d}V_\mathrm{g}$ and the mixing factor $\dot{\xi}_x\dot{\xi}_z \cos\phi$. Symmetric and asymmetric antennas made of multiple dipole blocks are designed to manipulate the local terahertz field distribution and near-field enhancement. Simulations of the terahertz antennas reveal that the self-mixing factor is greatly enhanced in the left and right edge areas of the gated electron channel. The phase flip of π of the perpendicular terahertz field across the gated electron channel induces a photocurrent in the left edge area opposite in direction to that induced within the right edge area. An asymmetric antenna design allows for optimization of the overall photocurrent according to the structural factor $\int_0^L \dot{\xi}_x\dot{\xi}_z \cos\phi \,\mathrm{d}x$.

For realistic design of an antenna coupled self-mixing detector, the influence of bonding pads on the antenna response and the near-field enhancement are evaluated. Meander-shaped low-pass filters are applied to effectively isolate the antenna blocks from the metallic bonding pads. In the next chapter, the physical realization of self-mixing terahertz detectors based on AlGaN/GaN field-effect transistors with terahertz antennas will be introduced in detail.

References

1. Dyakonov, M., Shur, M.S.: Shallow water analogy for a ballistic field effect transistor: new mechanism of plasma wave generation by dc current. Phys. Rev. Lett. **71**, 2465 (1993)
2. Dyakonov, M., Shur, M.S.: Detection, mixing, and frequency multiplication of terahertz radiation by two-dimensional electronic fluid. IEEE Trans. Electron Devices **43**(3), 380–387 (1996)
3. L ü, J.Q., Shur, M.S., Hesler, J.L., Liangquan, S., Weikle, R.: Terahertz detector utilizing two-dimensional electronic fluid. IEEE Electron Device Lett. **19**(10), 373–375 (1998)
4. L ü, J.Q., Shur, M.S.: Terahertz detection by high-electron-mobility transistor: enhancement by drain bias. Appl. Phys. Lett. **78**, 2587 (2001)
5. Tauk, R., Teppe, F., Boubanga, S., Coquillat, D., Knap, W., Meziani, Y.M., Gallon, C., Boeuf, F., Skotnicki, T., Fenouillet-Beranger, C., Maude, D.K., Rumyantseva, S., Shur, M.S.: Plasma wave detection of terahertz radiation by silicon field effects transistors: responsivity and noise equivalent power. Appl. Phys. Lett. **89**(25), 253511 (2006)
6. Knap, W., Dyakonov, M., Coquillat, D., Teppe, F., Dyakonova, N., Sakowski, J., Karpierz, K., Sakowicz, M., Valusis, G., Seliuta, D., Kasalynas, I., El Fatimy, A.: Field effect transistor for

terahertz detection: physics and first imaging applications. J. Infrared Millim. Terahz Waves **30**(12), 1319–1337 (2009)
7. Knap, W., Teppe, F., Meziani, Y., Dyakonova, N., Lusakowski, J., Buf, F., Skotnicki, T., Maude, D., Rumyantsev, S., Shur, M.S.: Plasma wave detection of sub-terahertz and terahertz radiation by silicon field-effect transistors. Appl. Phys. Lett. **85**, 675 (2004)
8. Lisauskas, A., Pfeiffer, U., Öjefors, E., Bolìvar, P.H., Glaab, D., Roskos, H.G.: Rational design of high-responsivity detectors of terahertz radiation based on distributed self-mixing in silicon field-effect transistors. J. Appl. Phys. **105**, 114511 (2009)
9. Knap, W., Rumyantsev, S., Lu, J., Shur, M., Saylor, C., Brunel, L.: Resonant detection of subterahertz radiation by plasma waves in a submicron field-effect transistor. Appl. Phys. Lett. **80**, 3433 (2002)
10. El Fatimy, A., Teppe, F., Dyakonova, N., Knap, W., Seliuta, D., Valuis, G., Shchepetov, A., Roelens, Y., Bollaert, S., Cappy, A., Rumyantsev, S.: Resonant and voltage-tunable terahertz detection in InGaAs/InP nanometer transistors. Appl. Phys. Lett. **89**, 131926 (2006)
11. Popov, V.V., Polischuk, O.V., Knap, W., El Fatimy, A.: Broadening of the plasmon resonance due to plasmon-plasmon intermode scattering in terahertz high-electron-mobility transistors. Appl. Phys. Lett. **93**, 263503 (2008)
12. Peralta, X.G., Allen, S.J., Wanke, M.C., Harff, N.E., Simmons, J.A., Lilly, M.P., Reno, J.L., Burke, P.J., Eisenstein, J.P.: Terahertz photoconductivity and plasmon modes in double-quantum-well field-effect transistors. Appl. Phys. Lett. **81**, 1627 (2002)
13. Dyer, G.C., Vinh, N.Q., Allen, S.J., Aizin, G.R., Mikalopas, J., Reno, J.L., Shaner, E.A.: A terahertz plasmon cavity detector. Appl. Phys. Lett. **97**, 193507 (2010)
14. Dyer, G.C., Aizin, G.R., Allen, S.J., Grine, A.D., Bethke, D., Reno, J.L., Shaner, E.A.: Induced transparency by coupling of Tamm and defect states in tunable terahertz plasmonic crystals. Nat. Photonics **7**, 925–930 (2013)
15. Veksler, D., Teppe, F., Dmitriev, A.P., Kachorovskii, V.Yu., Knap, W., Shur, M.S.: Detection of terahertz radiation in gated two-dimensional structures governed by dc current. Phys. Rev. B **73**, 125328 (2006)
16. Sun, J.D., Qin, H., Lewis, R.A., Sun, Y.F., Zhang, X.Y., Cai, Y., Wu, D.M., Zhang, B.S.: Probing and modelling the localized self-mixing in a GaN/AlGaN field-effect terahertz detector. Appl. Phys. Lett. **100**, 173513 (2012)
17. Sun, J.D., Sun, Y.F., Wu, D.M., Cai, Y., Qin, H., Zhang, B.S.: High-responsivity, low-noise, room-temperature, self-mixing terahertz detector realized using floating antennas on a GaN-based field-effect transistor. Appl. Phys. Lett. **100**, 013506 (2012)
18. Sun, Y.F., Sun, J.D., Zhou, Y., Tan, R.B., Zeng, C.H., Xue, W., Qin, H., Zhang, B.S., Wu, D.M.: Room temperature GaN/AlGaN self-mixing terahertz detector enhanced by resonant antennas. Appl. Phys. Lett. **98**, 252103 (2011)
19. Sun, J.D., Sun, Y.F., Zhou, Y., Zhang, Z.P., Lin, W.K., C.H., Zeng, Wu, D.M., Zhang, B.S., Qin, H., Li, L.L., Xu, W.: Enhancement of terahertz coupling efficiency by improved antenna design in GaN/AlGaN HEMT detectors. AIP Conf. Proc. **1399**, 893 (2011)
20. Zhou, Y., Sun, J.D., Sun, Y.F., Zhang, Z.P., Lin, W.K., Lou, H.X., Zeng, C.H., Lu, M., Cai, Y., Wu, D.M., Lou, S.T., Qin, H., Zhang, B.S.: Characterization of a room temperature terahertz detector based on a GaN/AlGaN HEMT. J. Semicond. **32**(4), 064005 (2011)
21. Sun, J.D., Qin, H., Lewis, R.A., Yang, X.X., Sun, Y.F., Zhang, Z.P., Li, X.X., Zhang, X.Y., Cai, Y., Wu, D.M., Zhang, B.S.: The effect of symmetry on resonant and nonresonant photoresponses in a field-effect terahertz detector. Appl. Phys. Lett. **106**, 031119 (2015)
22. Lü, L., Sun, J.D., Lewis, R.A., Sun, Y.F., Wu, D.M., Cai, Y., Qin, H.: Mapping an on-chip terahertz antenna by a scanning near-field probe and a fixed field-effect transistor. Chin. Phys. B **24**(2), 028504 (2015)
23. Brews, J.R.: A charge-sheet model of the MOSFET. Solid State Electron. **21**(2), 345–355 (1978)
24. Taflove, A., Hagness, S.C.: Computational Electrodynamics: the Finite-difference Time-domain Method, 3rd edn. Artech House, Boston (2005)
25. Yee, K.: Numerical solution of initial boundary value problems involving Maxwell's equations in isotropic media. IEEE Trans. Antennas Propag. **14**(2), 302–307 (1966)

Chapter 3
Realization of Terahertz Self-Mixing Detectors Based on AlGaN/GaN HEMT

Abstract In this chapter, the fabrication, characterization, and optimization of self-mixing terahertz field-effect detectors based on AlGaN/GaN 2DEG are introduced in details. By fabrication, five different detectors are made to uncover the self-mixing mechanism and search for an optimized detector design. By characterization, we not only obtain the $I - V$ characteristics, the responsivity, the noise-equivalent power, the response spectrum, the response speed, the polarization effect, etc, but also we probe the localized self-mixing photocurrent based on which the quasi-static detector model and the design of asymmetric antenna are verified. Under the guidance of the detector model, we focus on the design of terahertz antennas and field-effect gate to improve the detector responsivity and sensitivity. An asymmetric antenna with three dipole blocks is found to be the most effective antenna among the five different designs. A design rule for high-sensitivity terahertz detectors is given.

3.1 Introduction

3.1.1 Gallium Nitride and High-Electron-Mobility Transistors

In semiconductor industry, silicon and germanium are the first generation of semiconductors. Compound semiconductors of III-V GaAs, InP, GaP, InAs, and AlAs etc are the second generation of semiconductors. Semiconductors with the band gap $E_g > 2.3$ eV are the third generation of semiconductors including CdS (2.42 eV), SiC (3.2 eV), ZnO (3.32 eV), GaN (3.42 eV), ZnS (3.68 eV), AlN (6.20 eV), etc. Compared with previous two generations, the third generation has great development and application potential [1–3]. Therefore, it has become the research hotspot in semiconductor materials and devices all over the world. As the main representative of the third generation semiconductors, gallium nitride has the advantages of large band gap, high critical breakdown electric field, high thermal conductivity, high-electron saturation velocity and lower dielectric constant, resistance to radiation, and good chemical stability. Comparing to conventional metal oxide semiconductor field-effect transistors (MOSFETs), High-electron-mobility transistor (HEMTs) based on AlGaN/GaN

J. Sun, *Field-effect Self-mixing Terahertz Detectors*, Springer Theses,
DOI 10.1007/978-3-662-48681-8_3

heterostructures have a few advantages. Due to the strong spontaneous polarization, the 2DEG in AlGaN/GaN heterostructure has a high-electron density in the order of $10^{13}\,\text{cm}^{-2}$.

In the field of microelectronics and optoelectronics, high speed and/or high power are the main issues to concern. Important performance parameters of typical semiconductors including Si, GaAs, SiC, and GaN are shown in Table 3.1. It is obvious that the third generation semiconductors has advantages in large band gap, high critical breakdown field, high thermal conductivity, high-electron saturation velocity, and lower dielectric constant and etc. In the case of GaN, it is chemically stable, high temperature resistant, corrosion resistant, radiation harden, high frequency, and high power. It has shown many potential applications in the microwave, radio frequency, power electronics, and terahertz technology.

The Johnson quality factor (JMF) and the Baliga quality factor (BFOM), two key parameters to characterize a semiconductor material for high power and high frequency applications, can be expressed as

$$\text{JMF} = (E_\text{B} v_\text{s})^2 (4\pi^2). \tag{3.1}$$

$$\text{BFOM} = \varepsilon \mu_n E_\text{B}^2. \tag{3.2}$$

Table 3.1 Properties of Si, GaAs, SiC, and GaN

Parameters	Si	GaAs (AlGaAs/GaAs)	4H-SiC	GaN (AlGaN/GaN)
Band gap (E_g/eV)	1.11	1.43	3.2	3.4
Dielectric constant ε	11.4	13.1	9.7	9.8
Breakdown field $E_\text{B}/(\text{Vcm}^{-1})$	6×10^5	6.5×10^5	3.5×10^6	3.5×10^6
Electron saturation velocity	1×10^7	1×10^7	2×10^7	1×10^7
$v_\text{s}/(\text{cms}^{-1})$		(2.1×10^7)		(2.7×10^7)
Electron mobility	1500	8,500	700	900
$\mu_n/(\text{cm}^2\text{V}^{-1}\text{s}^{-1})$		(10,000)		(2,000)
Thermal conductivity $k/(\text{Wcm}^{-1}\text{K}^{-1})$	1.5	0.5	4.9	1.3
Maximum working temperature T/°C	175	175	650	600
JMF (Johnson quality factor)	1	5	136	246
BFOM (Baliga quality factor)	1	9	13.5	39

where ε is the dielectric constant, μ is the electron mobility, E_B is breakdown electric field, v_s is the electron saturation velocity. As shown in Table 3.1, GaN has the obvious advantages for high power and high frequency applications.

A heterostructure consists of an interface of two semiconductors with different band gaps. As shown in Fig. 3.1, AlGaN is the wide band gap semiconductor and GaN has a narrower band gap. For AlGaN, as shown in Fig. 3.1a, the conduction band minimum, the valence band maximum, the Fermi level, the semiconductor work function, the electron affinity, and the built-in voltage of AlGaN are notated as E_{C1}, E_{V1}, E_{F1}, ϕ_{S1}, χ_{S1}, and V_{B1}, respectively. Similarly, the conduction band minimum, the valence band maximum, the Fermi level, the semiconductor work function, the electron affinity, and the built-in voltage of GaN are labeled as E_{C2}, E_{V2}, E_{F2}, ϕ_{S2}, χ_{S2}, and V_{B2}, respectively. The energy discontinuities between the valence bands and the conduction bands are $\Delta E_V = E_{V1} - E_{V2}$ and $\Delta E_C = E_{C1} - E_{C2}$, respectively. The difference in Fermi level is $\Delta E_F = E_{F1} - E_{F2}$. As shown in Fig. 3.1b, the discontinuities in the energy bands lead to band bending, strong electric field in the interface of AlGaN/GaN and a triangular quantum well in GaN. In the quantum well, the Fermi level becomes higher than the conduction band. Electrons are accumulated in the discrete quantum states of the quantum well. When only few of the lowest

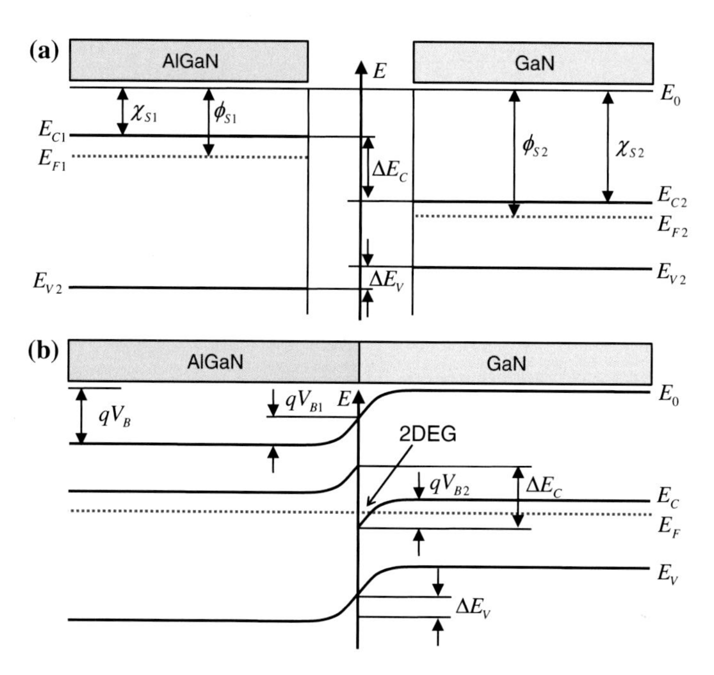

Fig. 3.1 Band diagrams of **a** independent AlGaN and GaN and **b** AlGaN/GaN heterostructure. E_C is the conduction band minimum, E_V is the valence band maximum, E_F is the Fermi level, ϕ_S is the work function, χ_S is the electron affinity and V_B is the built-in voltage. A 2DEG is confined in the triangular quantum well

quantum states are occupied, a quasi 2DEG is formed at the interface on the GaN side.

The band gap of AlGaN is closely related to the component of Al and can be expressed as

$$E_g(x) = xE_g(\text{AlN}) + (1-x)E_g(\text{GaN}), \tag{3.3}$$

where, x is the component of Al. The difference of the conduction bands ΔE_C can be written as

$$\Delta E_C = 0.7[E_g(x) - E_g(0)]. \tag{3.4}$$

$E_g(\text{AlN}) = 6.13\,\text{eV}$, $E_g(\text{GaN}) = 3.42\,\text{eV}$. In the case of 27 % of Al, i.e., $Al_{0.27}Ga_{0.73}N$, the band offset $\Delta E_c = 0.53\,\text{eV}$ contributes to the formation of a deep potential well in AlGaN/GaN heterostructure. Electrons fill only the first one or two lowest quantum levels of the potential well, the motion of electron is limited by the potential well in direction perpendicular to the interface and the charge distribution is shown in Fig. 3.2. Due to the spontaneous and piezoelectric polarizations, no intentional doping is required for AlGaN/GaN heterostructure. The electron density of an AlGaN/GaN 2DEG is in the order of $10^{13}\,\text{cm}^{-2}$, about 5 times of that in AlGaAs/GaAs 2DEG and the electron mobility can be as high as $\sim 2000\,\text{cm}^2/\text{Vs}$. HEMTs based on AlGaN/GaN 2DEG offer high transconductance, high saturation current, and high cut-off frequency.

Different from the infinite potential well, an AlGaN/GaN heterostructure forms a triangular potential well. In quantum mechanics, the partial energy of electrons moving perpendicular to the interface is quantized into discrete energy levels while the partial energy of electrons moving in the pane of the 2DEG is a continuum. Electrons in the quantum well can be described by the following Schrödinger equation

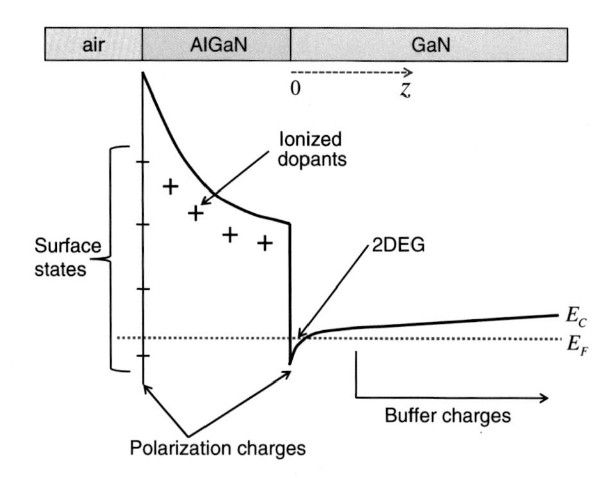

Fig. 3.2 Charge distribution in the AlGaN/GaN heterostructure

$$-\frac{\hbar^2}{2}\frac{\mathrm{d}}{\mathrm{d}z}\frac{\mathrm{d}}{m(z)\mathrm{d}z}\varphi + eV(z) + \Delta E(z) = E\varphi, \tag{3.5}$$

The boundary conditions of the triangular potential well are

$$V(z) = +\infty, \text{ when } z \leq 0, \tag{3.6}$$

$$V(z) = Fz, \text{ when } z \geq 0, \tag{3.7}$$

where, F is the built-in electric field intensity. Equation 3.5 can be solved as

$$\frac{\mathrm{d}^2\varphi(z)}{\mathrm{d}z^2} + \frac{2m^*}{\hbar^2}(E_i - qFz)\varphi(z) = 0, \tag{3.8}$$

where, m^* is the electron effective mass, q is the electron charge, E_i is the discrete energy levels. The energy of subband i can be expressed as

$$E_i(k) = E_i + \frac{\hbar k_x^2}{2m_x^*} + \frac{\hbar k_y^2}{2m_y^*}, \tag{3.9}$$

where, k_x and m_x^* are the wave vector and the electron effective mass in direction x, and k_y and m_y^* are the wave vector and the electron effective mass in direction y.

HEMTs are also called as heterostructure field-effect transistors (HFETs) or MODFETs. They are usually based on compound semiconductor heterostructure such as III-V including AlGaN/GaN, AlGaAs/GaAs, etc. Modern epitaxy techniques allow for high-quality epitaxial growth of AlGaN on undoped GaN buffer layer. HEMTs based on high-quality AlGaN/GaN heterostructures are becoming the next generation high-frequency and high power devices for radio-frequency and millimeter-wave applications. A schematic of an AlGaN/GaN HEMT is shown in Fig. 3.3a including a Schottky gate and the drain/source electrodes.

In thermal equilibrium, the band diagram along an intersection through the gated electron channel is shown in Fig. 3.3b. When a positive source–drain bias is applied, the potential drop along the source–drain connection leads to a variation of the band scheme in Fig. 3.3b parallel to the AlGaN/GaN interface. Depending on the local potential, the accumulation layer is more or less emptied of electrons; the position of the Fermi level E_F with respect to the band edges varies along the current channel. Transistor action is possible since an additionally applied gate voltage shifts the Fermi level in the gate metal with respect to its value deep into the undoped GaN layer (Fig. 3.3b). Most of the voltage drop occurs across this AlGaN layer, thus establishing a quasi-insulating barrier between the gate electrode and the 2DEG. The action of this Schottky barrier is similar to that of the SiO_2 layer in a metal oxide semiconductor FET (MOSFET). Depending on the gate voltage, the triangular potential well at the interface is raised or lowered in energy and the accumulation layer is emptied (Fig. 3.3c) or filled (Fig. 3.3b). This changes the carrier density of the 2DEG and switches the drain–source current. For large enough gate bias, the

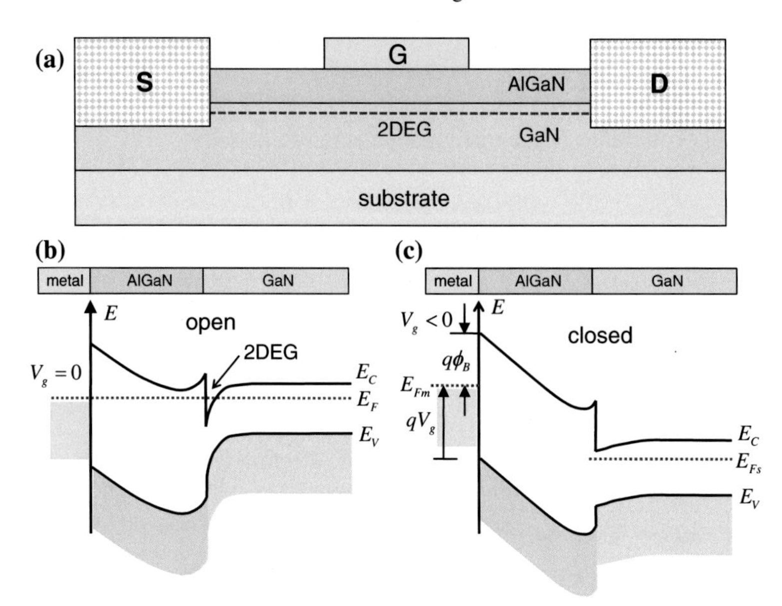

Fig. 3.3 **a** Schematic structure and **b**, **c** electronic band scheme perpendicular to the wafer surface underneath the gate electrode of a typical AlGaN/GaN HEMT device. **b** Under a zero gate voltage, the conductive channel is opened and formed by the 2DEG. **c** Under a negative gate voltage, the 2DEG are exhausted and the conductive channel is closed. Where, E_C is the conduction band minimum, E_V the valence band maximum, E_{Fm} the Fermi level in metal, E_{Fs} the Fermi level in the semiconductor, $q\Phi_B$ the Schottky barrier height, and V_g the gate voltage

depletion region penetrates into the 2DEG region, 2DEG concentrations become negligible and the channel is pinched off. The corresponding gate voltage is called the threshold voltage V_{th}.

A field-effect transistor can be approximately considered as a variable resistor tuned by the gate voltage. In the gradual channel approximation, the source–drain current can be expressed as

$$I = -en_s \upsilon W \tag{3.10}$$

where, n_s and υ are the charge density and the drift velocity of those free electrons in the gated electron channel, respectively. The charge density can be tuned by the gate voltage in the following way

$$n_s = C_g \left(V_g - V_{th}\right)/e, \tag{3.11}$$

where C_g the effective gate-channel capacitance per unit area. Under source–drain bias V_{ds}, the charge density at x in the channel can be expressed as

$$n_s(x) = C_g \left(V_g - V_x - V_{th}\right)/e. \tag{3.12}$$

The channel potential V_x varies from 0 to V_{ds} from $x = 0$ to $x = L$. According to Eq. (3.10), the source–drain current can be expressed as

$$I_{ds} = WC_g \left(V_g - V_x - V_{th}\right) \upsilon. \tag{3.13}$$

The minus sign is caused by the negative electron charge. The electron drift velocity υ can be expressed as $\upsilon = \mu E(x)$, where μ is the electron mobility and $E(x)$ the electric field in direction x. Due to $E(x) = -\mathrm{d}V/\mathrm{d}x$, the source–drain current can be expressed as

$$I_{ds} = \mu WC_g \left(V_g - V_x - V_{th}\right) \frac{\mathrm{d}V_x}{\mathrm{d}x}, \tag{3.14}$$

the source and drain potentials are $V(0) = 0\,\mathrm{V}$ and $V(L) = V_{ds}$, respectively. By multiplying $\mathrm{d}x$ to both sides of the above equation and integrating it through the gated electron channel,

$$\frac{1}{L}\int_{x=0}^{L} I_{ds}\,\mathrm{d}x = \frac{\mu WC_g}{L}\int_{V=0}^{V_{ds}} \left(V_g - V_x - V_{th}\right) \mathrm{d}V_x. \tag{3.15}$$

Because the current is all the same in a series circuit, therefore, the source–drain current can be written as

$$I_{ds} = \mu C_g \frac{W}{L}\left((V_g - V_{th})V_{ds} - \frac{1}{2}V_{ds}^2\right). \tag{3.16}$$

We can see that the channel current is controlled by both the source–drain voltage and the gate voltage. If the gate voltage is kept constant, the source–drain current as a function of the source–drain voltage is the output characteristic curve. Accordingly, if the source–drain voltage is kept constant, the channel current as a function of the gate voltage is the transfer characteristic curve. The output characteristic curve and the transfer characteristic curve of an AlGaN/GaN HEMT with a gate length of 2 μm and a gate width of 4 μm are shown in Fig. 3.4a and b, respectively. In the output characteristic curve, the channel current tends to saturate with the source–drain voltage reaches the saturation voltage V_{sat}. In the transfer characteristic curve, the channel current can be sensitively controlled by the gate voltage and pinched off at the threshold voltage $V_{th} = -4.1\,\mathrm{V}$ which is the threshold voltage.

The transconductance is a measure of the sensitivity of drain current density to the change in the gate voltage when the transistor is biased into the saturation regime ($V_{ds} > V_g - V_{th}$)

$$g_m = \frac{1}{W}\frac{\partial I_{ds}}{\partial V_g}\bigg|_{V_{ds}\equiv\mathrm{const.}}. \tag{3.17}$$

The transconductance is used for characterization the gain of the field-effect transistor. The transconductance as the function of the gate voltage can be measured, as shown in Fig. 3.5.

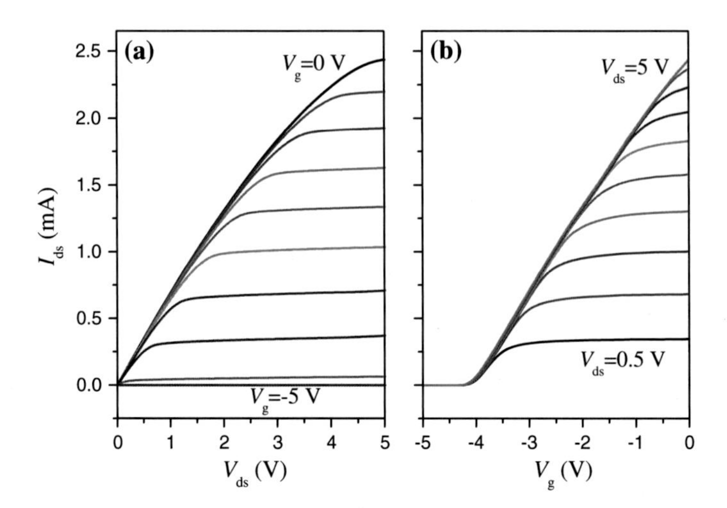

Fig. 3.4 **a** The output characteristic curve, **b** The transfer characteristic curve

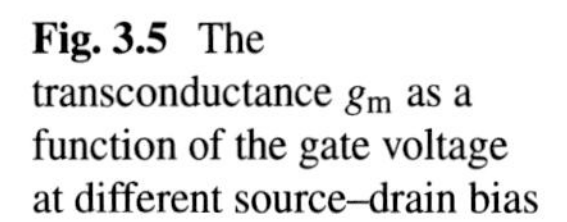
Fig. 3.5 The transconductance g_m as a function of the gate voltage at different source–drain bias

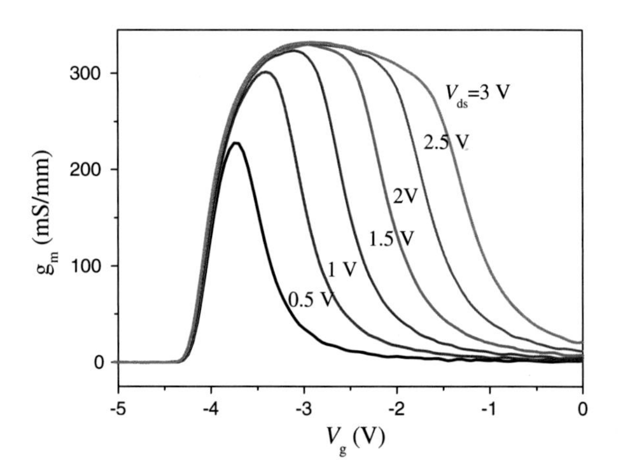

The regulation of the 2DEG density by the gate voltage can be regarded as the charging and discharging of the gate capacitor whose capacitance can be expressed as

$$C_g = \frac{\varepsilon_0 \varepsilon_s}{d}, \tag{3.18}$$

where ε_s is the dielectric constant of the semiconductor, ε_0 the dielectric constant of the vacuum and d is the effective distance between the gate and the 2DEG. The charge density n_s in the 2DEG is induced by the gate voltage V_g

$$C_g = e\frac{dn_s}{dV_g} \approx \frac{en_s}{V_g - V_{th}}. \tag{3.19}$$

When a HEMT is used in the saturation regime, the drain–source voltage is so high that the electrons in the channel move with the saturation velocity v_s ($\sim 10^7$ cm/s) independent of the drain–source bias. The drain–source current can be written as

$$I_{ds} \approx eWn_s\upsilon_s. \tag{3.20}$$

Evaluating the n_s from Eq. 3.19 and inserting it into Eq. 3.20, we have

$$I_{ds} \approx WC_g\upsilon_s\left(V_g - V_{th}\right). \tag{3.21}$$

At a constant drain–source voltage V_{ds}, the transconductance g_m can be simplified as

$$g_m = \frac{1}{W}\frac{\partial I_{ds}}{\partial V_g}\bigg|_{V_{ds}} \approx C_g\upsilon_s = e\upsilon_s\frac{dn_s}{dV_g}. \tag{3.22}$$

The transconductance is proportional to the field-effect factor dn_s/dV_g. As we have shown in the previous chapter, the self-mixing photocurrent is proportional to the field-effect factor and hence the transconductance. The transconductance of the HEMT device with a gate length and a width of 2 and 4 μm as a function of the gate voltage is shown in Fig. 3.5. The transconductance can be as high as 300 mS/mm which is essential to the detector responsivity and the sensitivity.

According to Eq. 3.22, the transconductance is proportional to the gate capacitance. The transit time τ for an electron to travel through the gated channel. The upper limit for the maximum frequency the field-effect can be operated

$$f_{max} = \frac{1}{\tau} = \frac{\upsilon_s}{L} = \frac{1}{L}\frac{g_m}{C_g}. \tag{3.23}$$

For a gate length of 1 μm and a saturation velocities of 10^7 cm/s, the maximum frequency is about 100 GHz. To improve the high frequency properties, the gate capacitance should be as low as possible and the transconductance as high as possible. It has to be mentioned that the maximum frequency f_{max} has nothing to do with the response frequency of the terahertz detectors. Also the above-mentioned transconductance as to be defined in the saturation regime, i.e., the source–drain bias is higher than the gate voltage swing ($V_{ds} > V_g - V_{th}$), does not apply to the detector's operation condition when a zero bias is applied. As we have shown in the previous chapter, the field-effect factor with and without the source–drain bias is $dG_0/dV_g = e\mu dn(V_g - V_{th})/dV_g$ and $dG/dV_{geff} = e\mu dn(V_g - V_{th} - V_x)/dV_{geff}$, respectively.

3.2 Detector Fabrication

In this section, we introduce the fabrication process of the terahertz detectors. To achieve high performance detectors, in addition to the rational design, state-of-art processing technologies are required, such as ultraviolet lithography (UVL), electron-beam lithography (EBL), electron-beam evaporation (EBE), atomic layer deposition (ALD), lift-off, inductively coupled-plasma etching (ICP), rapid thermal annealing (RTA), etc.

A typical antenna-coupled self-mixing terahertz detector based on an AlGaN/GaN 2DEG is shown in Fig. 3.6. The left shows the scanning-electron microscope image of the whole device and the right displays the zoom-in view of the central active region including the gate, the 2DEG channel/mesa, the terahertz antennas, and the low-pass filters.

The flow chart of the fabrication processes to make a field-effect detector based on an AlGaN/GaN 2DEG material is shown in Fig. 3.7. The main steps are wafer cleavage, wafer cleaning, isolation of the active region/mesa, gate dielectric layer deposition by ALD, formation of the ohmic contacts, formation of the gate and the filters, formation of the bonding pads, and packaging of the device. The details of each main step are introduced.

1. **Wafer cleavage and cleaning**. Free of contamination on the surface of the AlGaN/GaN 2DEG material is essential to make high performance detectors. The AlGaN/GaN heterostructure grown on sapphire substrate is divided into square-shaped samples with a dimension of 15 mm $\times$ 15 mm by laser cleavage. Organic contamination is removed by ultrasonic cleaning in acetone, isopropanol, ethanol, and deionized water. Removal of any ionic impurities on the sample surface is done in a mixed solution of ammonia and hydrogen peroxide.
2. **Formation of the active region/mesa**. The active region of the detector is in a form of AlGaN/GaN 2DEG mesa/channel. Photoresist is used as the mask for

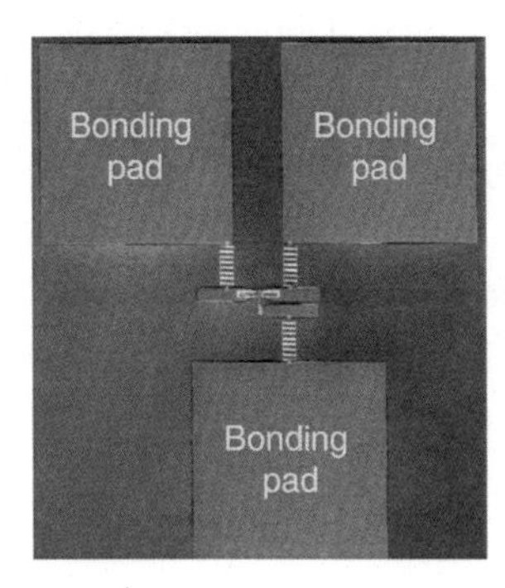

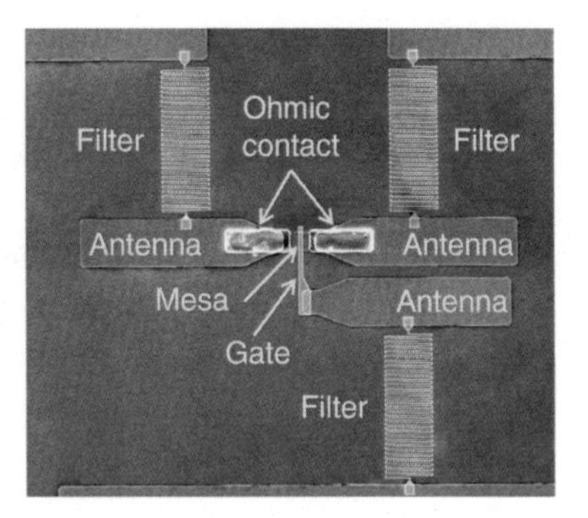

Fig. 3.6 Scanning-electron microscope images of a typical AlGaN/GaN field-effect terahertz detector. *Left* whole view of the detector with three bonding pads for the source, the drain, and the gate. *Right* zoom-in view of the central active region including the gate, the 2DEG channel/mesa, the terahertz antennas, and the low-pass filters

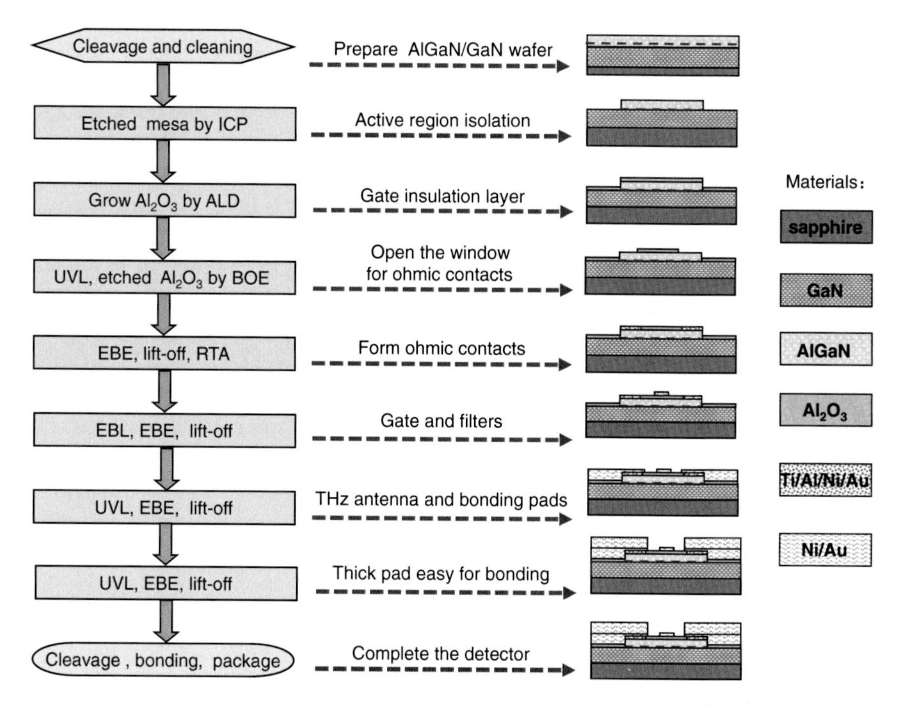

Fig. 3.7 Flow chart of the fabrication processes

etching away the unwanted portion of the 2DEG. An UVL is applied to define the geometry of the 2DEG mesa. Since there is no efficient chemical etchant for GaN and AlGaN, the etching of the unwanted 2DEG areas is done either by ion-beam etching (IBE) or reactive-ion-beam etching (RIE) or alternatively inductively coupled-plasma etching (ICP). Among them, IBE is a merely physical etching technology which decomposes GaN and AlGaN by ion bombardment. The etching rate of IBE is slow and the selectivity between GaN and AlGaN is relatively poor. Furthermore, the high-energy ions may degrade the 2DEG in terms of the electron mobility and the electron density. RIE uses a combined way of ion bombardment and chemical reaction. Compared to IBE, the etching rate and the etching selectivity of RIE are significantly improved. ICP is also based on ion bombardment and chemical reaction. But the selectivity and the etch rate are greatly improved. The plasma energy and density in ICP can be controlled to obtain a suitable etching rate and lower the damage to the semiconductors. In our process, both IBE and ICP have been tried and the ICP process yields a better result. The total depth of ICP eteching is about 50 nm which is about 25 nm deeper than the location of the 2DEG.

3. **Formation of the gate dielectric layer**. The gate leakage current is a serious problem in field-effect terahertz detector. After the formation of the 2DEG mesa and before the formation of the Schottky gate, a layer of dielectric material such as

Al_2O_3 may be deposited on the 2DEG material so that possible leakage current between the gate and the 2DEG channel can be minimized. As will be shown in the following sections, a negative gate voltage close to the pinch-off voltage $V_{th} \sim -4\,\text{V}$ is applied to the gate. Due to the thickness of the barrier in the AlGaN/GaN heterostructure is only about 23 nm, the electric field in the barrier is about $2 \times 10^6\,\text{V/cm}$. Defects and impurities in the AlGaN layer induce tunneling current which degrades the noise-equivalent power by lifting the noise current in the channel and by inducing charge fluctuation in the channel. In the worst case when the leakage current exceeds the photocurrent, the current preamplifier used to read out the photocurrent may be overload and the dynamic range is minimized. Therefore, reduction of the gate leakage current is a must to ensure a high sensitivity in field-effect terahertz detection.

However, a thick dielectric layer reduces significantly the gate capacitance and hence the field-effect factor $\mathrm{d}n_x/\mathrm{d}V_{\text{geff}}$ which is essential for high responsivity. Hence, unless the gate leakage current is the main concern, no dielectric layer will be used in the device fabrication. In case when a dielectric layer is necessary, the thickness has to be optimized to reach a balance between the leakage current suppression and the reduction in responsivity.

ALD has the advantages in preparing such a thin and smooth dielectric layers which has been applied for deposition of high-k dielectrics for CMOS technology. ALD is a method in which a film is grown on a substrate by exposing its surface to alternate gaseous species. The schematic diagram of ALD process is shown in Fig. 3.8. In contrast to chemical vapor deposition, different precursors are never present simultaneously in the reactor, instead they are introduced into the reactor one by one without overlap. In each introducing step, the precursor molecules react with the surface in a self-limiting way so that the reaction terminates by itself once all the reactive sites on the surface are consumed. The maximum amount of material deposited on the surface after a single exposure run of all

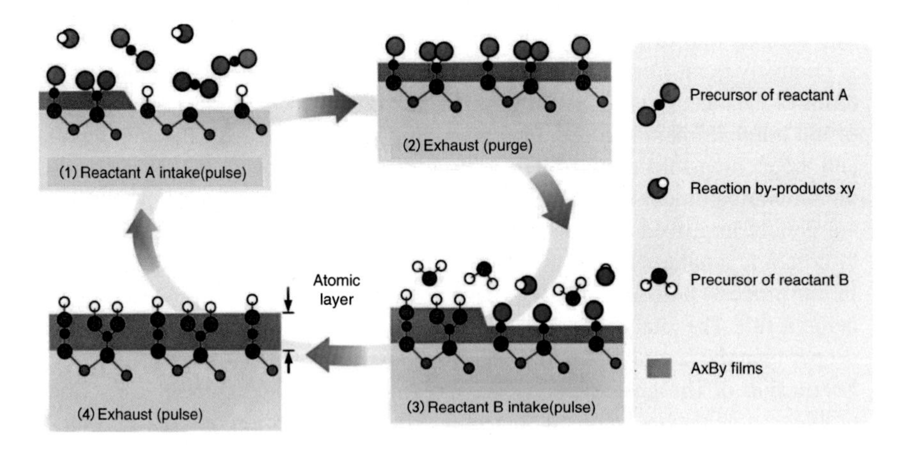

Fig. 3.8 Schematic diagram of atomic layer deposition (ALD)

of the precursors (an ALD cycle) is determined by the nature of the precursor-surface interactions [4]. By varying the number of cycles it is possible to grow materials precisely, uniformly and with high precision on arbitrarily complex and large-scale substrates. The advantages of ALD are the followings.

a. ALD provides a very controlled method in which an atomically precise film to the desired thickness can be deposited.
b. ALD allows for growth of multilayered dielectrics.
c. ALD runs at a relatively lower temperature comparing to other CVD processes. A lower temperature is beneficial when working with fragile substrates such as biological samples. Precursors that are thermally unstable may be used for ALD as long as their decomposition rate is slow.
d. Multicomponent nano layers can be deposited.
e. ALD is suitable for deposition on substrates with various shapes.

In our case, a thin layer of Al_2O_3 by ALD is deposited as the gate dielectric layer. By setting the thickness to be about 10 nm, the gate leakage current can be suppressed below picoampere and the mixing factor is maintained high.

4. **Formation of the ohmic contacts**. Two ohmic contacts serving as the source electrode and the drain electrode need to be formed on the 2DEG mesa to make electrical contacts to the electron channel. When a layer of Al_2O_3 is deposited on the 2DEG material, two windows need to be opened by removing the dielectric layer so that the metals for ohmic contacts can be deposited within. The windows are opened by an UV lithography and a following etching of Al_2O_3 in a buffered oxide etchant (BOE). A sequence of metal evaporation Ti/Al/Ni/Au= 80 nm/120 nm/70 nm/100 nm is done in an electron beam evaporator. A lift-off process in acetone removes the metal deposited out of the windows. In a RTA process at 850 °C with a forming gas of dry nitrogen, two ohmic contacts are formed in the window areas.
The process to make ohmic contacts is one of the key techniques of AlGaN/GaN HEMT manufacturing and determines the contact resistance which is essential to several important device parameters such as current density, gain, cut-off frequency, maximum operating temperature, high power performance, etc. A thin metal barrier compound with low resistivity, low work function, and good thermal stability must be formed to reduce the resistance of the ohmic contact for GaN and specifically AlGaN/GaN 2DEG. Ti, Ta, Zr, and Co meet the requirement. Titanium has a higher chemical activity and a low work function compared to several other metals. Therefore, Ti is the most commonly used ohmic contact metal. During the thermal annealing, TiN is formed in the reaction of Ti and N in AlGaN and a large number of donors are induced which turn the AlGaN layer into n^+ type. This process is eased and catalyzed by the second metal layer (Al) since aluminum is an ideal metal to enhance the solid phase reaction of N atoms with the Ti atoms. The third and fourth layers are the diffusion block layer and the cap layer, respectively. Due to the higher work function of the gold cap layer,

it is not suggested for the formation of the ohmic contact if the metal expansion to the semiconductor. Therefore, the diffusion block layer should be deposited between the cap layer and the barrier layer to prevent the mutual diffusion. In general, the metal with high melting point has good diffusion block property such as Pt, Pa, Ni, Cr, Mo, Ta, W, etc. The cap layer is used to maintain a stable and low-resistance contact to external electronics and to prevent the barrier layer from being oxidized. The general choice of the cap layer is the precious metal such as gold.

5. **Formation of Schottky gate, terahertz antennas, and/or terahertz filters**. The other key process in making high performance terahertz field-effect detectors is the preparation of the gate, the antennas, and/or terahertz filters which may be formed in one single step. Since the detector response is proportional to the field-effect factor $\mathrm{d}n_x/\mathrm{d}V_{\mathrm{geff}}$, it is preferred to make nanometer-sized Schottky gate by using the electron-beam lithography (EBL). Conventional UV lithography can be used to make Schottky gate, antennas and filters as well, however the minimum gate length would be around 1 μm. The metals for the gate, antennas, and filters are Ni/Au = 50 nm/200 nm. A lift-off process is required as usual.
6. **Bonding pads**. The final step to complete the fabrication of a detector is the formation of bonding pads. The bonding pads are made of Ni/Au = 50 nm/200 nm by electron-beam evaporation and lift-off on top of the ohmic contacts and the gate electrode.

In completion of the above processes, the detector chips are ready to be diced into single detectors with a dimension of about 3 mm $\times$ 3 mm by laser cleavage. A single chip will be fixed in a chip carrier and wire bonded to the pins of the carrier. A typical packaged device is shown in Fig. 3.9.

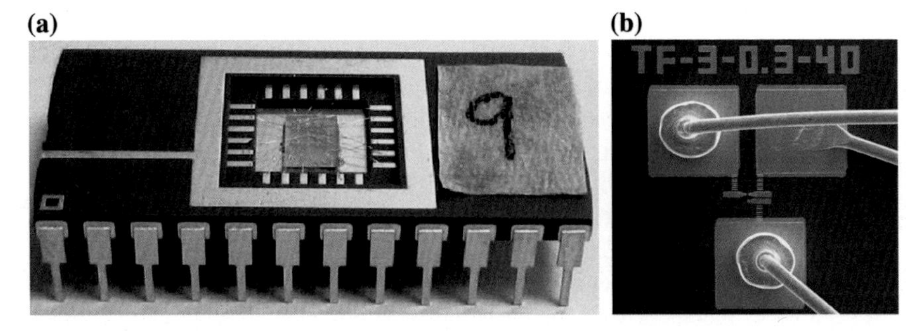

Fig. 3.9 **a** Optical microscope graph of a detector chip packaged in a DIP-24 chip carrier. **b** Scanning electron microscope graph of the detector showing the bonding wires and the bonding pads

3.3 Detector Characterization

In this section, detector characterization will be introduced. Through detailed electrical measurement in the DC limit and terahertz photocurrent/voltage measurement, the detectors are characterized in terms of the $I-V$ characteristics, gate voltage tuned terahertz photocurrent or photovoltage, optimal working gate voltage, responsivity, sensitivity, terahertz spectral response, polarization effect, response speed, etc. Most importantly, the detector physics and the detector model developed in Chap. 2 are studied by tuning the electron density in detectors with asymmetric and symmetric antennas. Based on these experiments and the verification of the detector model, we present different detector designs and demonstrate the detector optimization.

3.3.1 Terahertz Response at Zero Source–Drain Bias

In this section, we present high-responsivity terahertz self-mixing realized in an AlGaN/GaN HEMT detector [5]. We focus on the terahertz response at zero source–drain bias which is the most favorable operation mode for field-effect detectors [5–15]. Without a source–drain bias, i.e., there is no DC driving current flowing through the channel, shot noise, and the background current are eliminated. Also, the electron density in the channel can be tuned uniformly by the gate voltage. This becomes the most simple situation to understand the self-mixing mechanism in antenna-coupled field-effect detectors.

The device is fabricated on an AlGaN/GaN heterostructure which provides a 2DEG about 33 nm below the surface. The electron mobility and the density at 300 K are $\mu = 1{,}870\,\mathrm{cm}^2/\mathrm{Vs}$ and $n_s = 8.57 \times 10^{12}\,\mathrm{cm}^{-2}$, respectively. UVL rather than EBL was used to define the detector pattern. A partial top view of the device is shown in Fig. 3.10. The gate length is $L = 2\,\mu\mathrm{m}$ and the channel (mesa) width is $W = 8\,\mu\mathrm{m}$. Each antenna block is $45\,\mu\mathrm{m} \times 10\,\mu\mathrm{m}$. Only the lower right antenna block (g-antenna) is connected (to the gate). The gap between the gate and the two isolated blocks (i-antennas) is $1.5\,\mu\mathrm{m}$. The ohmic contacts are about $64\,\mu\mathrm{m}$ away from the gate. The total thickness of the device including the sapphire substrate is $416\,\mu\mathrm{m}$.

Upon terahertz irradiation with frequency $\omega = 2\pi f$ and power flux P_0, the induced photocurrent can be measured by using a current preamplifier which has a negligibly small input impedance ($r_{\mathrm{mc}} \approx 0\,\Omega$). The finite ohmic contact resistance and that from the 2DEG mesa in series with the gated channel are absorbed in $2r_1$. Based on the Eq. (2.39), the measured photocurrent can be rewritten as [5]

$$i = \frac{i_0}{1 + 2r_1 G_0} = \frac{1}{1 + 2r_1 G_0} \frac{\mathrm{d}G_0}{\mathrm{d}V_g} Z_0 P_0 \bar{z} \int_0^L \dot{\xi}_x \dot{\xi}_z \cos\phi \,\mathrm{d}x, \qquad (3.24)$$

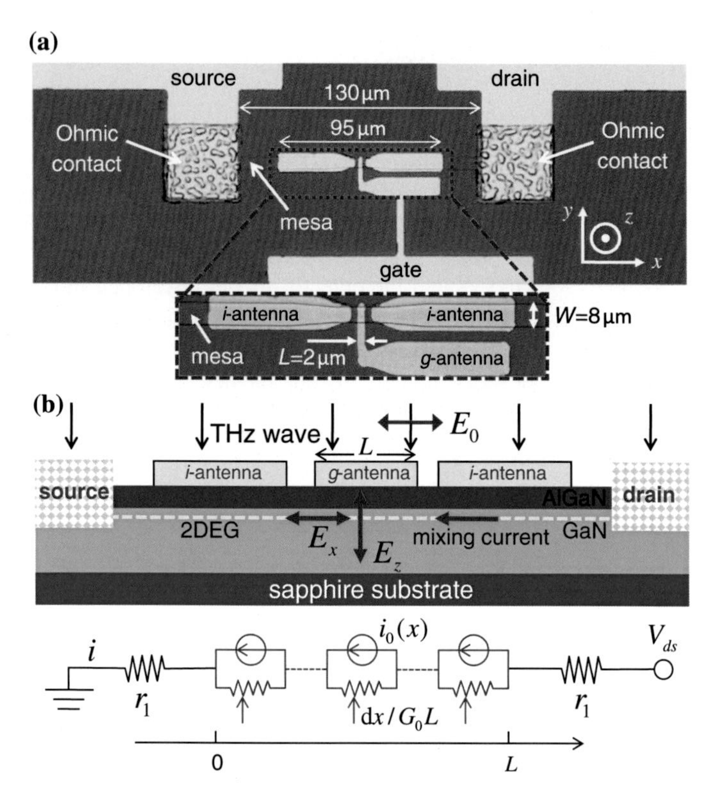

Fig. 3.10 Optical microscope image of the detector with an asymmetric antenna. *Inset* The central gate region showing the isolated antenna (i-antenna) structures

where i_0 is the internal mixing current, $G_0 = \mu C_g W(V_g - V_{th})/L$ is the channel conductance, C_g is the gate-channel capacitance per unit area, V_g is the applied DC gate voltage, and V_{th} is the threshold gate voltage. The integral in Eq. 3.24 represents the overall antenna enhancement, i.e., the antenna structural factor A.

In Fig. 3.11a, the simulated mixing factor $\dot{\xi}_x \dot{\xi}_z \cos \phi$ is plotted to reveal the spatial distribution at 900 GHz. As we have shown in Chap. 2 and Sect. 2.3.2, strong mixing occurs only at the edge of the gate. Due to the antenna asymmetry, the mixing at the left edges is about 4× stronger than that at the right edge. Furthermore, the mixing at the left edge generates a positive current opposite to that induced at the right edge. This inversion of polarity comes from a phase flip of π. This strongly unbalanced mixing is the origin of the observed unidirectional mixing current. According to Eq. 3.24, As shown in Fig. 3.11b, we find more than half of the mixing current is generated within 200 nm of the left side of the gated channel, even though the gate is 10 times longer. The rest of the gated channel simply acts as a resistor in series with the mixing part. The field-effect factor dG_0/dV_g is plotted in Fig. 3.11b to highlight the fact that there is an overlap between the mixing factor and the field-effect factor near the edges of the gated channel. A finite mixing current occurs only when the

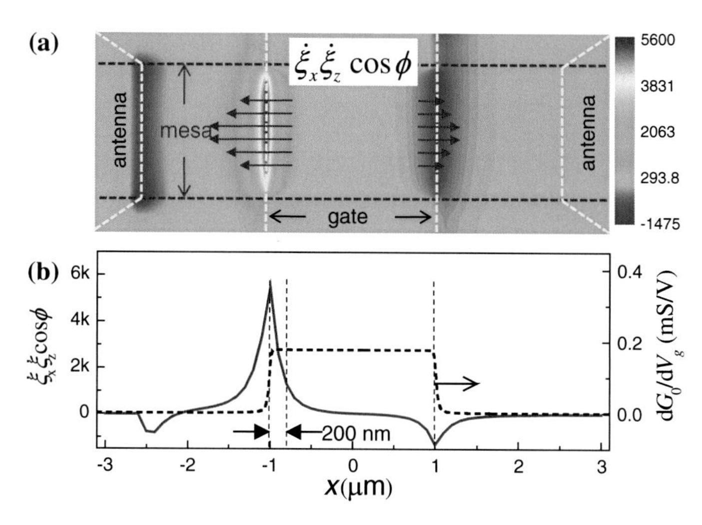

Fig. 3.11 Spatial distribution of the mixing factor from a FDTD simulation at 900 GHz. Reprinted with permission from Ref. [5], copyright 2012, American Institute of Physics

product of mixing fields and field-effect factor are nonzero. It can be seen that the device is an inherently nanometer-sized detector at terahertz frequencies.

The setup for characterizing the device is shown in Fig. 3.12a, where the terahertz radiation from a BWO terahertz source is chopped (f_m = 317 Hz), collected, collimated, and focused onto the detector located in a liquid nitrogen dewar. Two photos of the real setup are shown in Fig. B.1 in Appendix B. As the detector is insensitive to visible light, a 5 mm thick Polymethylpentene (TPX) disk is used as the window. For calibration, the terahertz beam is split by a high-resistivity silicon wafer (BS) and

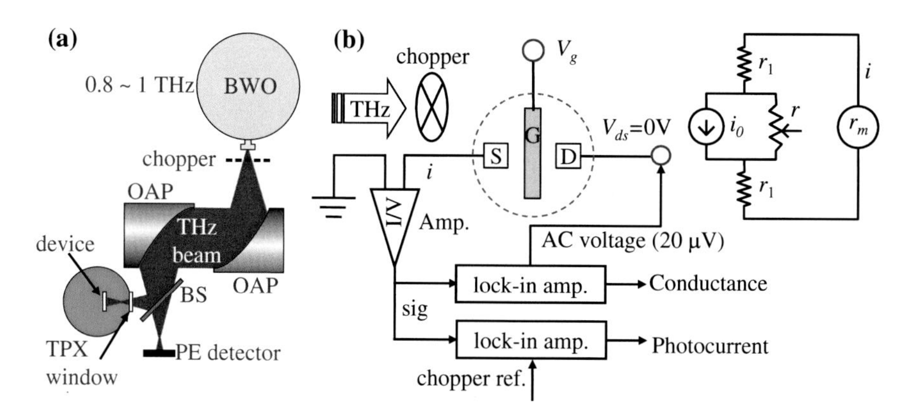

Fig. 3.12 **a** Schematic of the experiment setup. **b** Scheme of dual lock-in technique to measure the differential conductance and the photocurrent simultaneously. The *inset* is an equivalent circuit for detector including the series resistance (r_1) and the input resistance (r_m) of the current preamplifier

guided to a pyroelectric detector (Model SPI-A-62THZ from the Spectrum Detector Inc.). As shown in Fig. 3.12b, the detector is operated as a photocurrent sensor with zero source–drain bias. Only a small AC bias ($f_{AC} = 37$ Hz, 20 μV) is applied to the drain. Both the modulated AC source–drain current and the induced photocurrent are amplified by a DL1211 current preamplifier. Since the terahertz light and the source–drain bias are modulated at different frequencies ($f_{AC} < f_m$), a dual lock-in technique following the same current amplifier is applied to extract the photocurrent (i_T) and the differential conductance ($G = dI_{ds}/dV_{ds}$) simultaneously.

The source–drain conductance G and its derivative dG/dV_g are measured at 300 and 77 K, as shown in Fig. 3.13a. The conductances can be controlled by the gate voltage sensitively and the pinch-off voltage is $V_g = -4$ V. The derivative reaches the maximum value at 300 K and 77 K at $V_g = -3.68$ V and $V_g = -3.85$ V, respectively. Upon terahertz irradiation with $f = 897$ GHz and an estimated power $P_0 = 48$ nW, the mixing current as a function of the gate voltage is shown in Fig. 3.13b. The corresponding maximum current responsivity (R_i) is estimated to be 71 mA/W and 3.6 A/W at 300 K and 77 K, respectively. Shown as the solid lines in Fig. 3.13b, the fitting curves based on Eq. 3.24 are in good agreement with the experimental data.

The detector performance is greatly enhanced at lower temperatures. The extracted internal conductance G_0 and its derivative dG_0/dV_g are plotted in Fig. 3.13c. The derivative is increased by a factor of 2 and 3.5 at 300 K and 77 K, respectively. The

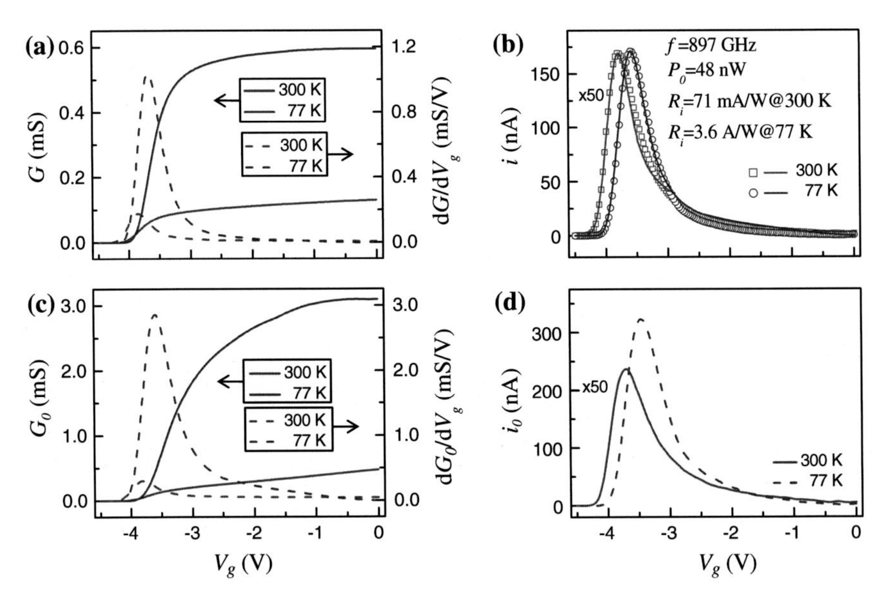

Fig. 3.13 Self-mixing characteristics at 300 and 77 K, different device parameters are given as a function of the gate voltage. **a** The measured source–drain conductance and its derivative. **b** The measured photocurrent. **c** The deduced internal conductance and its derivative. **d** The deduced internal photocurrent. Reprinted with permission from Ref. [5], copyright 2012, American Institute of Physics

internal photocurrent i_0 is increased by a factor of 1.5 and 2 at 300 K and 77 K, as shown in Fig. 3.13d, respectively. The extracted circuit parameters are summarized in Table 3.2. An enhancement factor of 51 in photocurrent at 77 K relative to 300 K is observed. This enhancement is attributed to the combined effect of enhanced electron mobility, a decrease in series mesa resistance and an enhanced antenna efficiency: $0.80 \times 9.22 \times 7.2 \approx 51$. The increase in antenna efficiency ($\times 7.2$) is a result of the higher conductivity of gold and lower damping of GaN at 77 K.

Not only the temperature affects the responsivity, but also the bonding pads, ohmic contacts and in fact any metallic structures nearby the terahertz antenna strongly influence the terahertz response. For comparison, we simulated a control detector (CD) fabricated on the same substrate with the same dimensions for the antenna and the gate, but, unlike in the original detector (OD), with the upper antenna blocks connected to the source and drain contacts. In Fig. 3.14a, the simulated mixing factor $\dot{\xi}_x \dot{\xi}_z \cos \phi$ is plotted and compared to the original design. The self-mixing factor of the control device is decreased by 40 %.

In Fig. 3.14c, d, the measured conductance, its derivative and the responsivity are compared to the control detector. The circuit parameters of both devices are listed in Table 3.3. Although the product of the scaling factor and the derivative of the conductance is about $0.96 \times 0.7 \approx 67\,\%$ of the control detector, the observed responsivity is enhanced by a factor of 9.1. This enhancement comes from the boost in antenna efficiency which is deduced to be $\times 15$. However, as shown in Fig. 3.14a the FDTD simulation (without taking into account the realistic ohmic properties) suggests that the floating antennas provide an enhancement factor of only $\times 1.5$. The enhancement mechanism is similar to that reported earlier using meander filters (see Ref. [6] and Sect. 3.4). The strong discrepancy (a factor of $15/1.5 = 10$) between the experimental data and the simulation indicates that the terahertz field is reduced by a factor of 10 due to the existence of the local ohmic contacts (less conductive and rougher than pure gold antennas) in the control detector. This cannot be easily modeled in FDTD since the nature of the annealing process at high temperature, the uniformity, and reproducibility of ohmic contacts are less controllable than those floating gold antennas made in a standard lift-off process. The floating antenna design effectively avoids this difficulty and maintains the high efficiency of the original antenna.

The detector can be operated to output a photovoltage instead of a photocurrent. The open-circuit photovoltage is measured by a voltage preamplifier which has an input impedance of 100 MΩ. The measured photovoltage as a function of the gate voltage at 897 GHz is shown in Fig. 3.15. The maximum voltage responses are about 3.6 kV/W and 33.6 kV/W at 300 K and 77 K, respectively. The best working points of the gate voltage are about $V_g = -4.03$ V and $V_g = -3.95$ V at 300 K and 77 K. Comparing to those for the optimal current responsivity, they are more closer to the threshold voltage. The photovoltage and the photocurrent under the same gate voltage obey the Ohm's theorem

$$v = i/G. \tag{3.25}$$

Table 3.2 Measured and fitting parameters at 300 and 77 K

T (K)	μ (cm^2/Vs)	G_0 (mS)	$2r_1$ (kΩ)	$(1+2r_1G_0)^{-1}$ –	dG_0/dV_g (mS/V)	Integral (10^{-6} V^2)	R_i (mA/W)	NEP ($pW/\sqrt{Hz}$)
77	1.58×10^4	4.8×10^{-1}	1.4	0.60	2.86	109	3600	2
300	1.87×10^3	5.9×10^{-2}	5.8	0.75	0.31	15	71	40
		8.5	8.1	0.80	×9.22	×7.2	≈ 51	

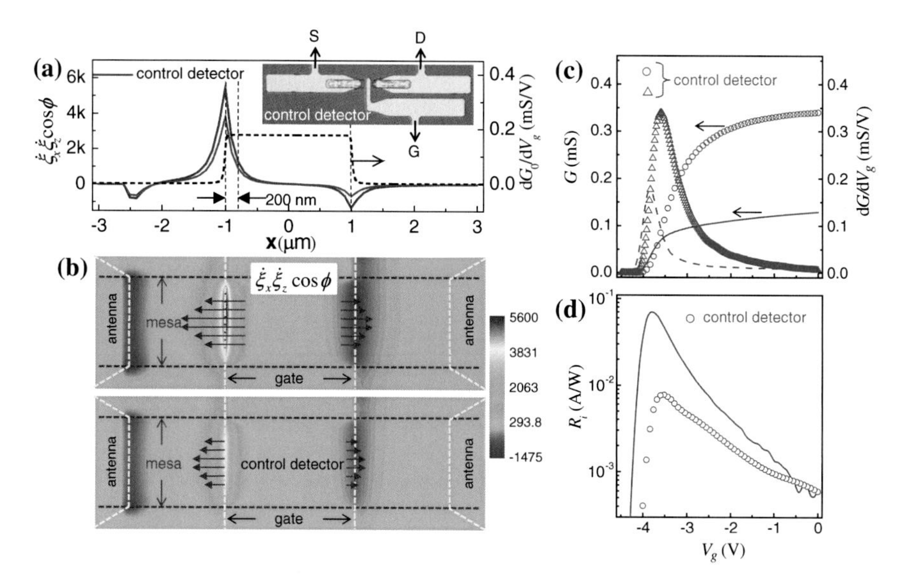

Fig. 3.14 Comparison between the original device and the control device. **a** Line plots of the mixing factor as a function of the location along the channel. **b** Two-dimensional color-scale plot of the mixing factor for the original detector (*upper panel*) and the control detector (*lower panel*). **c** Measured conductance and the derivative at 300 K. **d** Measured terahertz responsivities for both devices at 300 K. Reprinted with permission from Ref. [5], copyright 2012, American Institute of Physics

Table 3.3 Measured and extracted parameters from the original detector and the control detector

Detector	G_0 (mS)	$2r_1$ (kΩ)	$(1+2r_1G_0)^{-1}$ –	$\mathrm{d}G_0/\mathrm{d}V_g$ (mS/V)	Integral (10^{-6} V^2)	R_i (mA/W)	*NEP* (pW/$\sqrt{\text{Hz}}$)
OD	5.9×10^{-2}	5.8	0.75	0.31	15	71	40
CD	1.2×10^{-1}	1.2	0.78	0.44	1	7.8	500
			0.96	×0.70	×15	≈9.1	

The conductance and the measured photocurrent as a function of the gate voltage are shown in Fig. 3.16a, b, respectively. The calculated photovoltage based on Eq. (3.25) and the measured photocurrent shown in Fig. 3.16b is in a good agreement with the measured photovoltage as shown in Fig. 3.16c. The maximum in photovoltage does not mean a highest sensitivity, i.e., a lowest *NEP*, since the detector resistance and hence the thermal noise become much higher.

To summarize this section, we have the following conclusions.

1. The quasi-static self-mixing theory well describes the detector characteristics under a zero source–drain bias.
2. The terahertz response is closely related to the derivative of the conductance or the field-effect factor $\mathrm{d}G/\mathrm{d}V_g$.

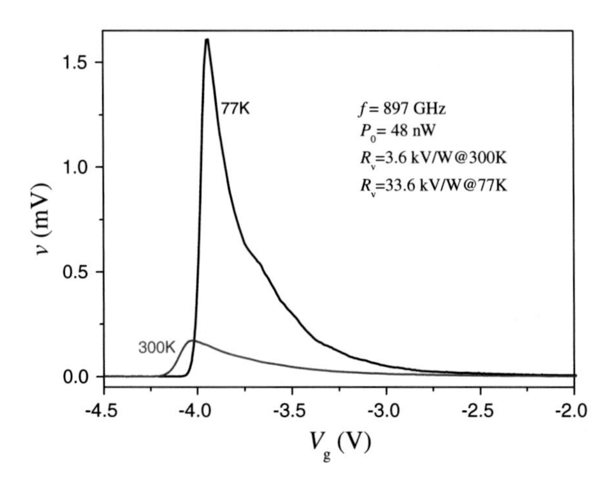

Fig. 3.15 Measured terahertz photovoltage as a function of the gate voltage at 897 GHz

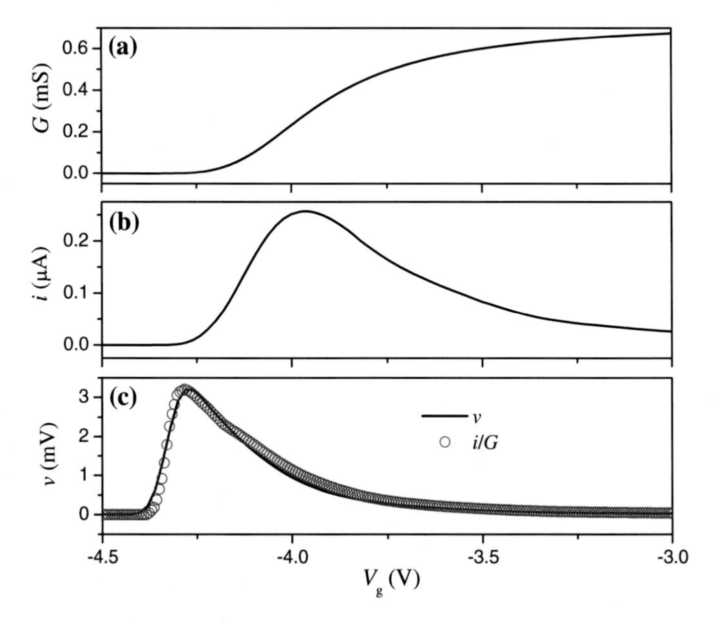

Fig. 3.16 **a** The detector conductance, **b** the photocurrent, and **c** the photovoltage as a function of the gate voltage. The *solid curve* in (**c**) is calculated from the conductance curve in (**a**) and the photocurrent curve in (**b**). Gate voltages for the optimal responsivities in the current mode and the voltage mode are significantly different, i.e., the detector impedance and hence the noise are different as well

3. Asymmetric antenna greatly enhances the self-mixing factor $\dot{\xi}_x\dot{\xi}_z \cos\phi$ and so as to improve the terahertz responsivity.
4. Antenna efficiency is influenced by bonding pads and ohmic contacts nearby the antennas.
5. The optimal gate voltages for the maximum responsivity in current mode and voltage mode are different.

3.3.2 Terahertz Response Under a Finite Source–Drain Bias

In Sect. 3.3.1, we have presented the characteristics of the terahertz responses under zero source–drain bias. The quasi-static self-mixing theory well describes the detector characteristics. The theory based on shallow water model in Sect. 2.1 did not account for the electron velocity saturation and is applicable only to the linear region [16, 17]. However, few direct evidence has been obtained on the terahertz field distribution and the validity of the quasi-static model. In this section, direct evidences can be found by characterizing the terahertz response under a finite source–drain bias. By tuning the source–drain bias and the gate voltage, the distribution of the electron density and the location-dependent field-effect factor $\mathrm{d}n_x/\mathrm{d}V_{\mathrm{geff}}$ can be tuned. In the following, a new measurement setup and the terahertz response as a function of the source–drain bias and the gate voltage will be presented. The effect of the symmetries in the spatial distributions of the electron density and the terahertz electric field is revealed experimentally for the first time.

The unique measurement setup applied to reveal the physics of self-mixing is shown in Fig. 3.17. The device under test is the same as that has been discussed in Sect. 3.3.1 [5]. A scanning-electron micrograph of the detector is embedded in Fig. 3.17 to illustrate the measurement scheme. The terahertz wave, generated from the same BWO terahertz source ($f = \omega/2\pi = 903\,\mathrm{GHz}$), is collimated and focused onto the detector cooled at 77 K. The device is cooled at 77 K so that the signal-to-noise ratio of the photocurrent is enhanced. We use a standard lock-in technique, in which the terahertz wave is modulated by a chopper at frequency $f_{\mathrm{m}} = 317\,\mathrm{Hz}$. The directional photocurrent is calculated from the magnitude and the phase of

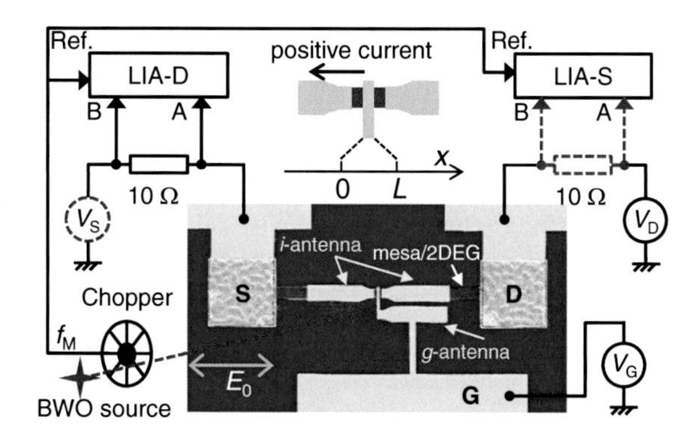

Fig. 3.17 Measurement circuit diagram. The inset is an artificially colored scanning-electron micrograph of the detector. Reprinted with permission from Ref. [18], copyright 2012, American Institute of Physics

the AC voltage across a sampling resistor (10 Ω). The DC voltage source (V_S or V_D) and the sampling resistor are connected to the drain electrode and the source electrode, respectively, or vice versa. The order of the lock-in amplifier's (LIA-S or LIA-D) differential inputs (A and B) is kept consistent with the direction of a positive photocurrent.

Again, the detector is first characterized without terahertz irradiation. As shown in Fig. 3.18a, the current-voltage curves exhibit the standard characteristics of a field-effect transistor in the linear regime (LR: $V_{ds} < V_g - V_{th}$) which can be described by

$$i_x = e\mu W n_x \frac{dV_x}{dx}, \tag{3.26}$$

where, the channel potential V_x varies from 0 to V_{ds} from $x = 0$ to $x = L$. The local electron density is controlled by the effective local gate voltage: $n_x = C_g V_{geff}/e = C_g(V_g - V_{th} - V_x)/e$, where C_g is the effective gate-channel capacitance per unit area. By measuring the differential conductance ($G_0 = dI_{ds}/dV_{ds} \approx \mu C_g W(V_g - V_{th})/L$) at $V_{ds} = 0$ V as shown in Fig. 3.18b, the threshold voltage and the gate capacitance are estimated to be $V_{th} \approx -4.34$ V and $C_g \approx 0.4\,\mu\text{F/cm}^2$, respectively. We can clearly distinguish the linear regime, the saturation regime, the transition regime and the pinch off regime from Fig. 3.18b. In the static GCA model, the source–drain current and the local channel potential can be written as

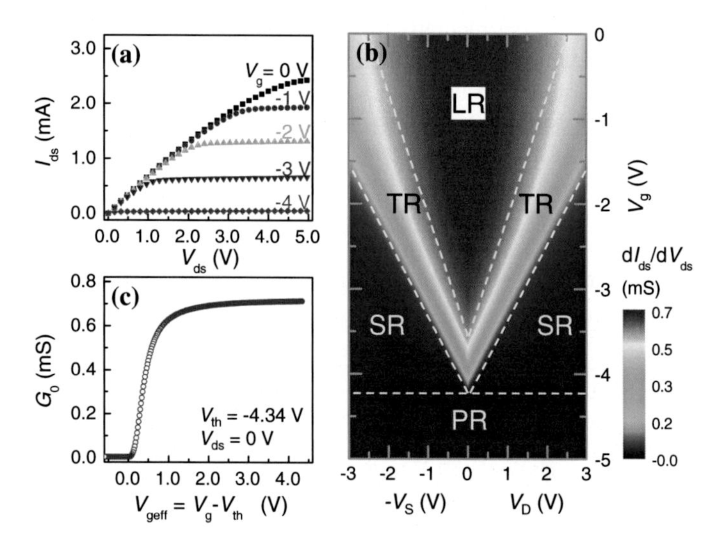

Fig. 3.18 **a** The $I - V$ characteristics of the detector. **b** The differential conductance at $V_{ds} = 0$ V as a function of V_g. **c** A color-scale plot of the differential conductance as a function of V_g and drain/source bias (V_D or V_S) which is applied either at the drain or at the source. The *dashed lines* separate the map into the linear regime (LR), the transition regime (TR), the saturation regime (SR), and the pinch-off regime (PR). Reprinted with permission from Ref. [18], copyright 2012, American Institute of Physics

$$I_{ds} = \begin{cases} \mu C_g W\left[2(V_g - V_{th})V_{ds} - V_{ds}^2\right]/2L & \text{if } V_{ds} \le V_g - V_{th}, \\ \mu C_g W\left[(V_g - V_{th})^2 + \lambda(V_{ds} - V_g + V_{th})\right]/2L & \text{if } V_{ds} > V_g - V_{th}, \end{cases} \quad (3.27)$$

$$V_x = \begin{cases} (V_g - V_{th})\left[1 - (1 - x/L_{LR})^{1/2}\right], x = [0, L] & \text{if } V_{ds} \le V_g - V_{th}, \\ (V_g - V_{th})\left[1 - (1 - x/L_{SR})^{1/2}\right], x = [0, L_{SR}] & \text{if } V_{ds} > V_g - V_{th}, \end{cases} \quad (3.28)$$

where, parameter $\lambda \approx 0.08$ V describes the degree of the effective channel-length modulation. $L \rightarrow L_{SR} = L/[1 + \lambda(V_{ds} - V_g + V_{th})/(V_g - V_{th})^2]$ in regime SR, and $L_{LR} = L(V_g - V_{th})^2/[2(V_g - V_{th})V_{ds} - V_{ds}^2]$ in regime LR. From Eq. 3.28 and the experimental data shown in Fig. 3.18b, the charge density and its derivative $\mathrm{d}n/\mathrm{d}V_{geff}$ at different locations are numerically calculated.

Similar to that shown in Sect. 3.3.1, upon terahertz irradiation with a frequency of ω and an energy flux of P_0, both a horizontal ($\dot{\xi}_x E_0$) and a perpendicular ($\dot{\xi}_z E_0$) terahertz field in the channel are induced. As shown in Fig. 3.19a, b from a FDTD simulation, the horizontal field is concentrated in the gaps between the gate and the i-antennas, while the perpendicular field is mainly distributed under the gate. Both fields are stronger at the source side of the gate than at the drain side. The horizontal field vanishes and changes its phase by π at $x_C \approx +0.3\,\mu$m, while the perpendicular field keeps its phase constant along the channel (Fig. 3.19c, d). The

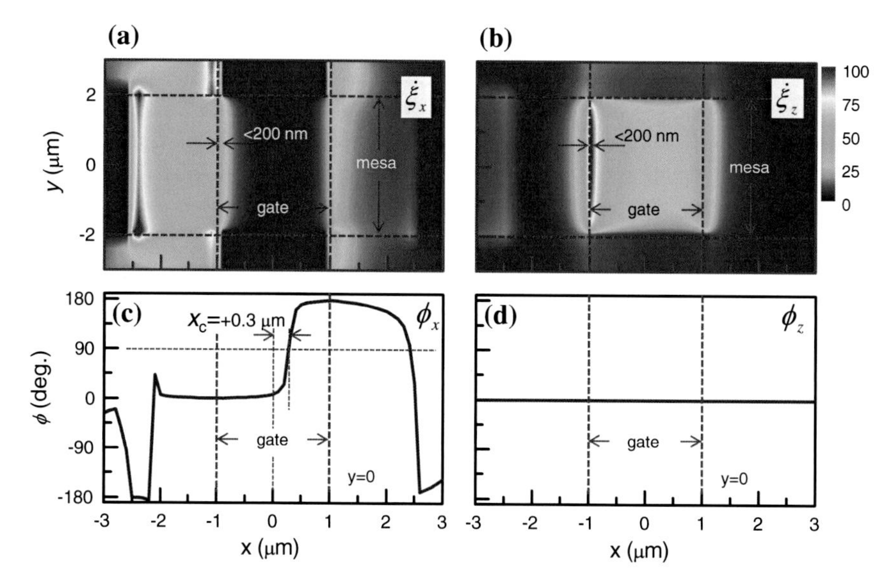

Fig. 3.19 Spatial distributions of the field enhancement factors (**a**, **b**) and phases (**c**, **d**) of the horizontal and the perpendicular field at 900 GHz. Reprinted with permission from Ref. [18], copyright 2012, American Institute of Physics

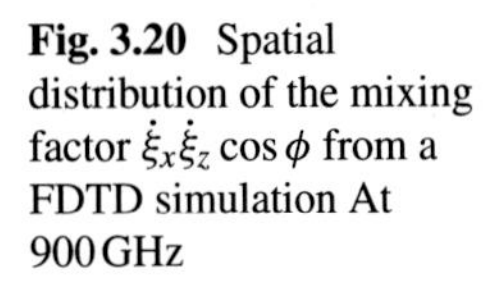
Fig. 3.20 Spatial distribution of the mixing factor $\dot{\xi}_x\dot{\xi}_z \cos\phi$ from a FDTD simulation At 900 GHz

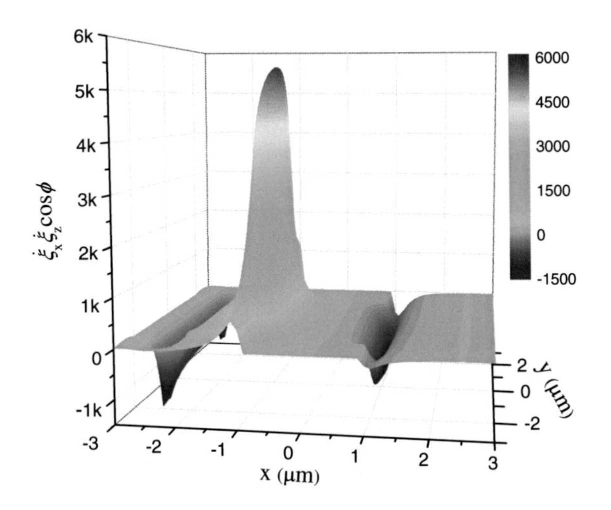

spatial distribution of the mixing factor $\dot{\xi}_x\dot{\xi}_z \cos\phi$ obtained from a FDTD simulation is shown in Fig. 3.20. The asymmetric antennas induce rather localized (<200 nm) terahertz fields. Due to the antenna asymmetry, the mixing factor at the left edge is about $\times 4$ stronger than that at the right edge. Because of a phase flip of π, the mixing at the left edge $x = -1\,\mu$m generates a positive current about 6,000, opposite to that induced at the right edge $x = 1\,\mu$m about $-1{,}500$. This strongly unbalanced mixing give rise to the observed unidirectional photocurrent.

At 77 K, the momentum relaxation time is about $\tau \approx 2$ ps, which corresponds to a mean-free-path of $\lesssim 1\,\mu$m. In spite of the fact that $\omega\tau \approx 10$, in a nanometer-sized transistor, the electron transit time ($\tau_D \ll 200\,\text{nm}/v_D \approx 2$ ps) and the plasma-wave transit time ($\tau_p \ll 200\,\text{nm}/s \approx 0.2$ ps) become smaller than the period of the oscillating field, i.e., $\omega\tau_{D/p} \ll 10$. In the following, we will present the map of terahertz response in the 2D space of V_{ds} and V_{g} and verify the quasi-static self-mixing model taking into account the nonuniform terahertz excitation. According to the model developed in Chap. 2, the terahertz photocurrent is

$$\begin{aligned} i_{\text{T}} &= i_{xz} + i_{xx} \\ &= \frac{e\mu W}{2L} Z_0 P_0 \int_0^L \frac{\text{d}n}{\text{d}V_{\text{GE}}} \left[\dot{\xi}_x\dot{\xi}_z \cos\phi - \dot{\xi}_x\xi_x\right] \text{d}x. \end{aligned}$$

where $\phi = \phi_x$ since $\phi_z = 0$. Current i_{xz} is induced by both horizontal and perpendicular fields, while i_{xx} is only from the horizontal field. For numerical simulation, we approximate the term i_{xz} as

$$i_{xz} = \frac{e\mu W}{2L} Z_0 P_0 \bar{z} \int_0^L \frac{\text{d}n}{\text{d}V_{\text{GE}}} \dot{\xi}_x\dot{\xi}_z \cos\phi \, \text{d}x, \tag{3.29}$$

where, $\bar{z} = \xi_z/\dot{\xi}$ is the effective distance between the channel and the gate. Due to the phase change at x_C, mixing term i_{xz} changes from a positive value at the left side to a negative value at the right side of x_C. In the case of the antenna design shown in Fig. 3.19, the terahertz potential $\xi_x E_0$ is obtained by integrating the horizontal field from x_C to $x = 0$ and from x_C to $x = L$ by setting $\xi_x(x_C) = 0$. The ratio of the horizontal field to the perpendicular field ranges from 0.6 at $x = 0$ and 0.5 at $x = L$ to 0 at $x = x_C$ with a mean value of about 0.19. Hence, the self-mixing from term i_{xx} is a minor contribution.

In Fig. 3.21a, the photocurrent as a function of $V_{D/S}$ and V_g is mapped in a color-scale plot. For comparison, a simulated map based on the above quasi-static model is shown in Fig. 3.21b. In the simulation, we used two sets of input data. One of them includes the parameters such as the electron mobility, the gate geometry, the threshold voltage, and the gate-controlled electron density obtained under zero bias as shown in Fig. 3.18c. The other set of input data includes the spatial distribution of the terahertz fields obtained from FDTD simulations, as shown in Fig. 3.20. The simulation reproduces the main features of the polarity change and the magnitude variation. In regime PR, the whole channel is pinched off and the terahertz mixing is disabled. In the linear regime, there is only a very weak photocurrent. A strong photocurrent is produced in regime TR and SR, and strongly depends on the applied bias and gate voltage. When the bias is applied at the drain, a negative photocurrent

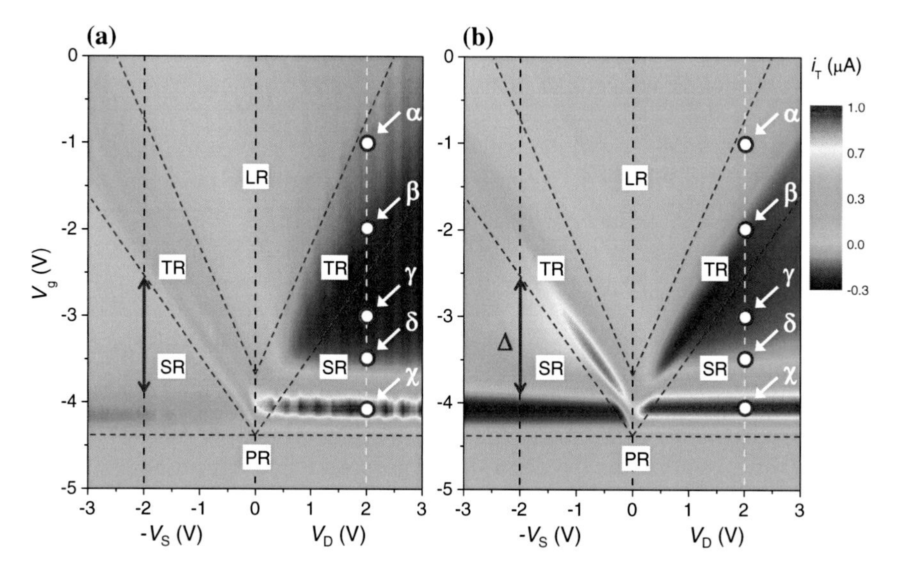

Fig. 3.21 The measured (**a**) and the simulated (**b**) terahertz response as a function of $V_{D/S}$ and V_g. Three vertical lines are marked at $V_S = 2$ V, $V_{D/S} = 0$ V, and $V_D = 2$ V. Along the line at $V_D = 2$, five bias conditions are marked by α, β, γ, δ and χ, corresponding to $V_g = -1.0, -2.0, -3.0, -3.5$, and -4.1 V, respectively. In the SR regime when the bias is applied at the source, there is a clear deviation between the experiment and the simulation. Reprinted with permission from Ref. [18], copyright 2012, American Institute of Physics

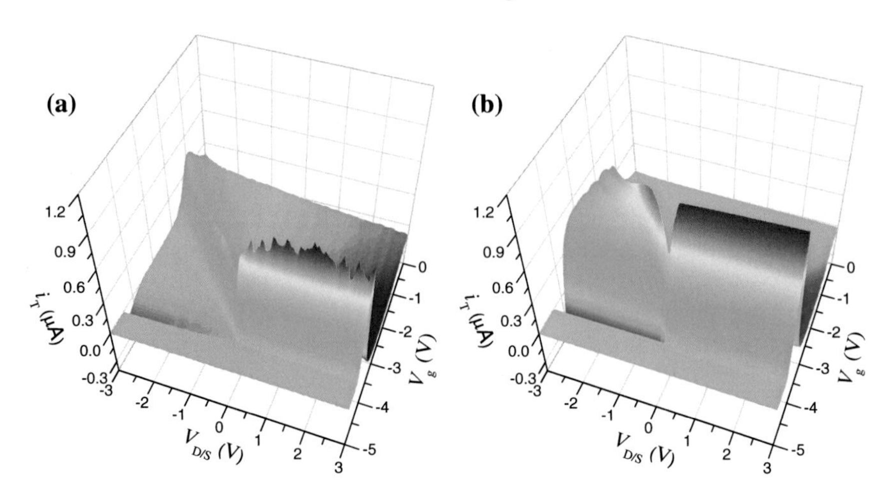

Fig. 3.22 Photocurrent as a function of $V_{D/S}$ and V_g plotted in 3D maps

is flipped into a stronger positive photocurrent when the gate voltage is swept from regime TR into regime PR. In contrast, when the bias is applied at the source and the gate voltage is swept, a positive current is flipped into a weaker negative current. When the bias is swept and switched to either the drain or the source with a constant gate voltage at −4.1 V, the photocurrent differs in polarity and its magnitude becomes fairly independent of the bias voltage. In order to show clearly the changing of polarities, we plot the terahertz response as a function of $V_{D/S}$ and V_g in a 3D map as shown in Fig. 3.22.

The spatial distributions of the terahertz fields ($\dot{\xi}_x \dot{\xi}_z \cos\phi$) and the field-effect factor (dn/dV_{GE}) at $V_g = -4.1$ V and $V_{D/S} = 2.0$ V are plotted in Fig. 3.23a. Due to the diminishing of charge modulation at the side where a positive bias is applied, the local self-mixing is suppressed. Hence, a polarity change is observed between the cases of $V_D = 2$ V and $V_S = 2$ V. To further illustrate the terahertz responses in different regimes, we mark three lines ($V_{D/S} = 0,\ 2$ V) on the maps shown in Fig. 3.21. Along the line at $V_D = 2$ V, we select five conditions labeled as α, β, γ, δ and χ, corresponding to $V_G = -1.0,\ -2.0,\ -3.0,\ -3.5$, and −4.1 V, respectively. The location-dependent charge modulation is plotted in Fig. 3.23b. By decreasing the gate voltage from α to χ, the charge modulation becomes stronger and the location of the maximum shifts from the drain side to the source side. This evolution gives rise to the observed transition from a weak negative current to a strong positive current along the line at $V_D = 2.0$ V. In Fig. 3.23c–e, the experimental gate-controlled photocurrent is compared to the simulation. At zero bias, the photocurrent resembles what we observed. At $V_D = 2.0$ V, the simulation fits well to the experimental data. As shown in Fig. 3.23e, when the bias is applied at the source side, a transition from positive current at $V_g \approx -2$ V to a weaker negative current at $V_g \approx -4.1$ V is reproduced. There is, however, a strong suppression of photocurrent from $V_g = -3.5$ V to $V_g = -2.5$ V (also marked as vertical double arrows in Fig. 3.21), while according

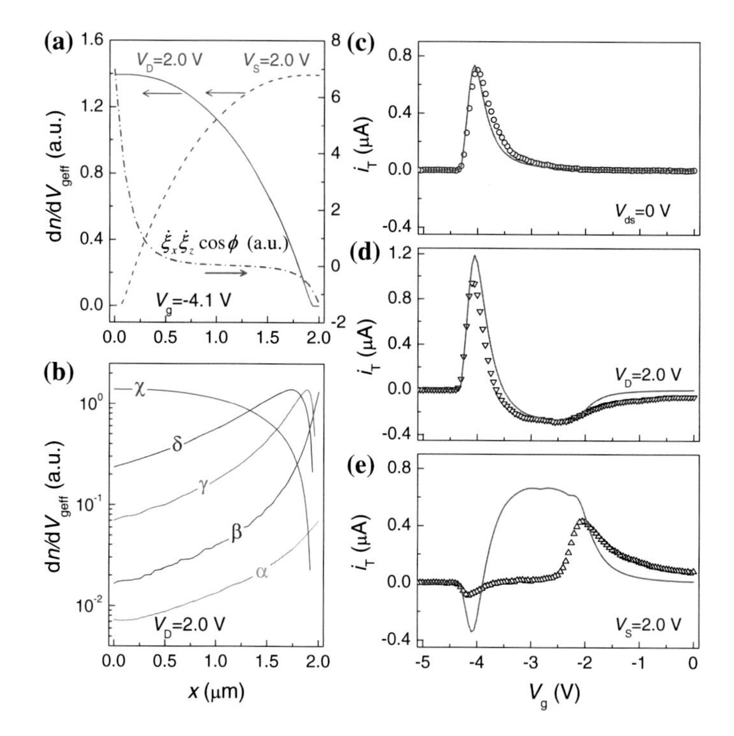

Fig. 3.23 **a** The spatial distributions of the charge modulation $\mathrm{d}n/\mathrm{d}V_{\mathrm{geff}}$ and the self-mixing term $\dot{\xi}_x\dot{\xi}_z \cos\phi$ at $V_g = -4.1\,\mathrm{V}$ with a DC bias of $2\,\mathrm{V}$ applied at the source or at the drain. **b** The spatial distributions of charge modulation at different gate voltages and at a constant drain bias ($V_{\mathrm{D}} = 2.0\,\mathrm{V}$). The mixing current as a function of V_G with **c** $V_{\mathrm{D/S}} = 0\,\mathrm{V}$, **d** $V_{\mathrm{D}} = 2.0\,\mathrm{V}$, and **e** $V_{\mathrm{S}} = 2.0\,\mathrm{V}$. Reprinted with permission from Ref. [18], copyright 2012, American Institute of Physics

to the quasi-static model a strong positive current should occur. The absence of photocurrent in region SR is also observed in other detectors with symmetric antennas. The common feature is that the self-mixing in regime SR is suppressed as long as the electron density at the same side with strong mixing factor is set to be mostly depleted. The deviation between the experiment and the model indicates that there is a different transport mechanism. The most likely candidate is the plasma-wave-induced photocurrent. To reveal the physics behind, systematic experiments at lower temperatures on different antenna designs to tune both the amplitude and the phase of the terahertz field are required. Accordingly, a time-dependent detector model taking into account the localized terahertz excitation will help to uncover the mystery.

It is shown in Fig. 3.23d that the current responsivity is enhanced by a positive source–drain bias applied to the drain. However, the measured noise current as a function of $V_{\mathrm{D/S}}$ and V_g shown in Fig. 3.24 exhibits a stronger noise in the saturation regime than that in the other regimes. A field-effect detector has a better sensitivity in the linear regime.

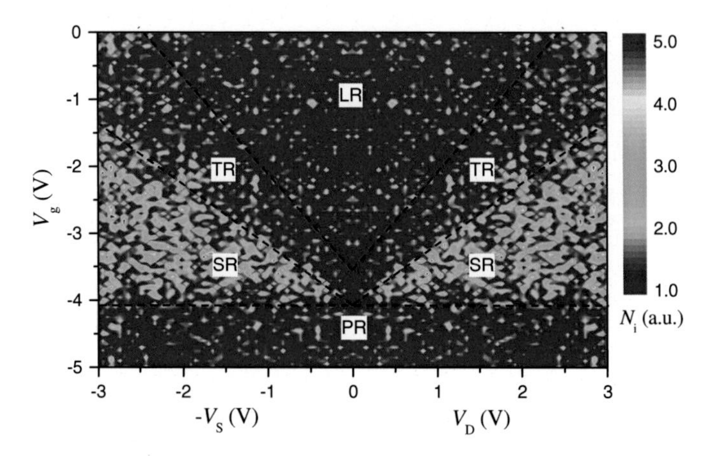

Fig. 3.24 The measured noise current as a function of $V_{D/S}$ and V_g

We can draw the following conclusions from the above experiment results.

1. The quasi-static self-mixing model is verified by comparing the simulated results with the experimental data in a two-dimensional space of the gate voltage and the source–drain bias.
2. The model well describes the main features of the terahertz response in the linear regime, the transition regime, the saturation regime, and the pinch-off regime.
3. The model well describes not only the magnitude of, but also the polarity of the photocurrent.
4. The simulated spatial distribution of the terahertz field and its phase for the specific terahertz antenna is confirmed.
5. It is preferable to operate the detector in the linear regime for high sensitivity.
6. It has to be mentioned that similar results were observed in other detectors with different antennas or a different channel length, such as nanometer-sized FETs and a quantum-point-contact device with a symmetric bow-tie antenna.
7. The deviation between the experiment and the model at one side of the saturation regime suggests that a full model taking into account the dynamic transport and the spatial distribution of terahertz excitation is desirable.

3.3.3 Responsivity and Noise-Equivalent Power

In this section, the responsivity and the *NEP*, the two most important figure-of-merits to characterize the detector performance, are discussed. According to the different signal readout methods, the responsivity can be written in the current responsivity and the voltage responsivity. The current responsivity can be expressed as [19, 20]

$$R_i = \frac{i}{P_0}, \tag{3.30}$$

where i is the photocurrent, P_0 is the effective power received by the detector. In fact, the radiation beam spot area S_t is much bigger than the active area S_a of the detector, therefore the current responsivity should be expressed as

$$R_i = \frac{iS_t}{P_t S_a}, \tag{3.31}$$

where P_t is the total power of the source. We defined the radiation beam spot area $S_t = \pi d^2/4$, where d is the radius the radiation beam spot. Since the area of the detector with the antenna is smaller than the diffraction limited area $S_\lambda = \lambda^2/4$, the active area S_a is taken as S_λ. Similarly, the voltage responsivity can be expressed as

$$R_v = \frac{vS_t}{P_t S_a}, \tag{3.32}$$

where v is the photovoltage.

Under a radiation at 901 GHz, the responsivity as a function of the gate voltage is characterized at 300 and 77 K, as shown in Fig. 3.25. The half-wavelength corresponding to this frequency is about 166 μm. Therefore, we defined the active area S_a of the detector is taken to be 200 μm × 200 μm, which is about 20 times greater than the effective area of the detector antenna 100 μm × 20 μm. The effective power P_0 is about 58 nW. The current responsivity as a function of the gate voltage is shown in Fig. 3.25a. The corresponding maximum current responsivity (R_i) is estimated to be 83 mA/W at 300 K. As the temperature is reduced to 77 K, benefited from the increase of the electron mobility and antenna coupling efficiency, the corresponding maximum current responsivity (R_i) is nearly increased by 50 times (about 4.1 A/W).

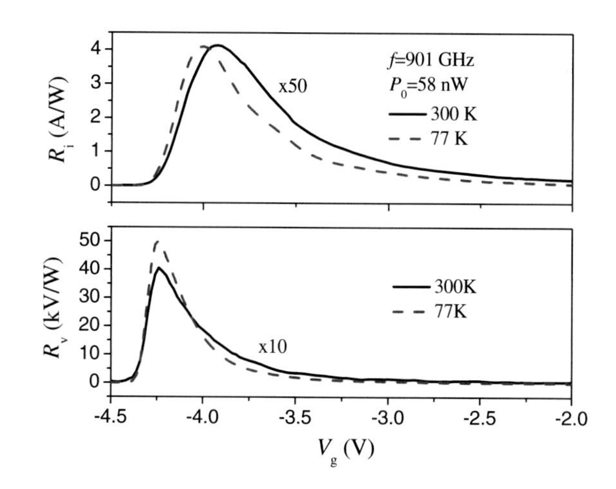

Fig. 3.25 Responsivity measured for 901 GHz as a function of the gate voltage at 300 and 77 K. **a** Current responsivity. **b** Voltage responsivity

The voltage responsivity as a function of the gate voltage is shown in Fig. 3.25b. The corresponding maximum voltage responsivity (R_v) is 4 kV/W and 50 kV/W.

The noise-equivalent power can be expressed as

$$NEP = \frac{N}{R}, \tag{3.33}$$

where N is the noise spectral density and R is the responsivity. The thermal noise current spectral density as the main noise source of the detector can be expressed by the channel conductance G as [21]

$$N_i = \sqrt{4k_B TG}. \tag{3.34}$$

Therefore, the noise-equivalent power limited by the thermal noise current spectral density can be expressed as

$$NEP = \frac{N_i}{R_i}. \tag{3.35}$$

For detection in the voltage mode, the thermal noise voltage spectral density can be expressed by the channel conductance G as [21]

$$N_v = \sqrt{4k_B T/G}. \tag{3.36}$$

The noise-equivalent power can be expressed as

$$NEP = \frac{N_v}{R_v}. \tag{3.37}$$

The measured noise-equivalent power can be defined as

$$NEP = \frac{N_{iB}}{R_i}. \tag{3.38}$$

The measured channel conductance controlled by the gate voltage at 300 and 77 K is shown in Fig. 3.26a. As the temperature reduced to 77 K, the conductance is increased by about 6 times. Through Eq. (3.34), we can get the current noise spectral density as shown in Fig. 3.26b at 300 and 77 K. With the gate voltage is reduced to the threshold voltage, the current noise spectral density decreased rapidly. The current noise spectral density is about 1 pA/$\sqrt{\text{Hz}}$ with the maximum current responsivity at $V_g \approx -4.0$ V. The noise spectral density and hence the *NEP* can be effectively reduced by decreasing the temperature.

The measured and calculated *NEP* with a bandwidth of 1.25 Hz as a function of the gate voltage are shown in Fig. 3.27. The minimum *NEP* is as low as 40 pW/$\sqrt{\text{Hz}}$ and 2 pW/$\sqrt{\text{Hz}}$ at 300 K and 77 K, respectively. In comparison, the thermal-noise limited *NEP* within the measurement bandwidth is about 10 pW/$\sqrt{\text{Hz}}$ at 300 K and 0.6 pW/$\sqrt{\text{Hz}}$ at 77 K. The measured *NEP* is only $\times 4$ the thermal-noise limited *NEP*.

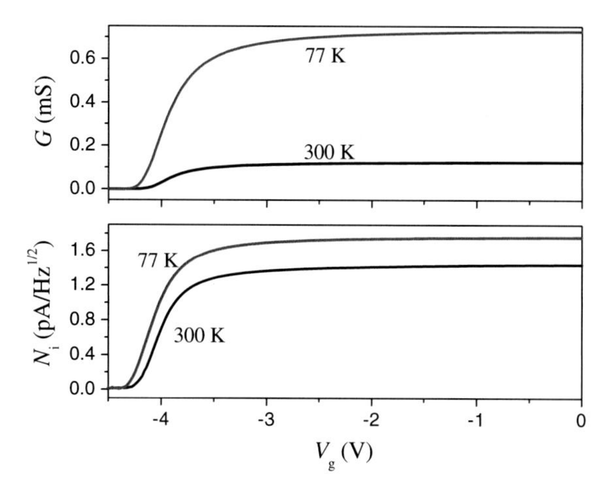

Fig. 3.26 **a** The measured channel conductance and **b** the current noise density spectrum as a function of the gate voltage at 300 and 77 K

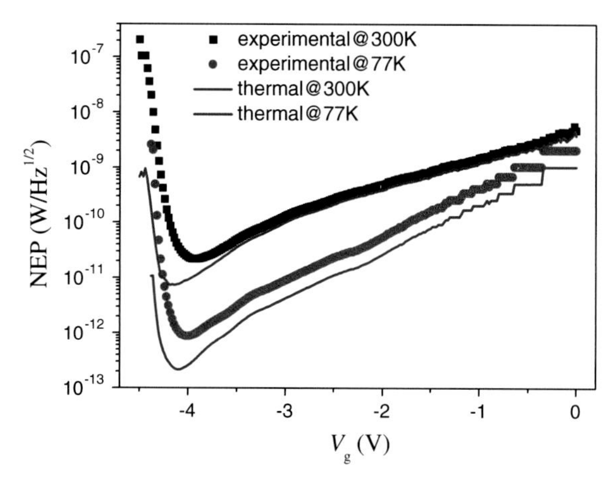

Fig. 3.27 The measured and calculated *NEP* as a function of the gate voltage at 300 and 77 K

The room-temperature *NEP* is comparable to that of a typical zero-biased Schottky-diode detector [22], and is better than Pyroelectric detectors and Golay detectors [23, 24]. It has to be noted that a similar *NEP* has been achieved in field-effect transistors based on silicon CMOS technology [25–27], although the operation frequency is around 650 GHz and 300 GHz. According to the model, *NEP* can be further reduced.

In the above experiments, the modulated terahertz photocurrnet is amplified by a current preamplifier (1×10^8 V/A) and followed by a lock-in amplifier which has a measurement bandwidth of 1.25 Hz. Hence, it is an indirect measurement of the noise. A signal analyzer would be the right equipment to measure the noise property of the detector in a wide frequency range. In a measurement, where the lock-in amplifier is replaced by a noise spectrum analyzer with an input impedance 50 Ω and a sampling bandwidth of 1 Hz, the measured terahertz responses and the noise density spectrum are shown in Fig. 3.28. The amplitude of the terahertz photocurrent signal keeps fairly constant when the modulation frequency is varied. The noise

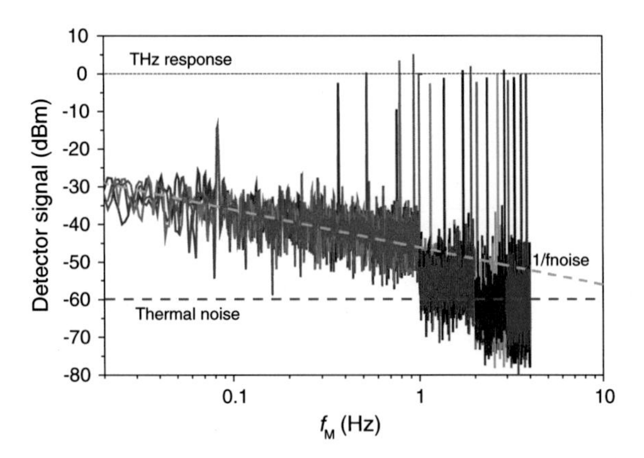

Fig. 3.28 The output signal from the detector in dBm measured by the signal analyzer. The terahertz power and hence the terahertz photocurrent are modulated at different frequencies

density quickly rolls off when the frequency increases and the $1/f$ noise is evident. A noise floor can be reached when the frequency is above 4 kHz. When terahertz detection is made at a modulation frequency above 4 kHz, the best signal-to-noise ratio as high as 10^4 can be achieved. By increasing the modulation frequency, the $1/f$ noise can be suppressed and the thermal noise becomes the dominant noise source.

The signal shown in Fig. 3.28 can be converted into the photocurrent signals by

$$i = \frac{\sqrt{1 \times 10^{-3} \times 10^{D/10} \times Z_s}}{Gain}, \tag{3.39}$$

where D is the detector signal (in dBm), $Z_s = 50\,\Omega$ is the the input impedance of the signal analyzer and *Gain* is the gain of the amplifier (in V/A), respectively. The photocurrent and the responsivity are shown in Fig. 3.29a and b, respectively. The photocurrent is constant about 3 nA at different modulation frequencies. The sensitivity *NEP* can be calculated directly from Fig. 3.28 according to

$$NEP = \frac{P_0}{\sqrt{B}} 10^{-SNR/20}, \tag{3.40}$$

where P_0 in W is the incident terahertz power received by the detector, *SNR* in dBm and B in Hz are the signal-to-noise ratio and the sampling bandwidth, respectively. The measured current noise is higher than the calculated thermal current noise which indicates excess noise is introduced by the measurement setup.

3.3.4 *Response Speed*

The response speed of a terahertz detector is of great importance in many terahertz applications where detection/sensing has to be made as fast as possible. Unlike

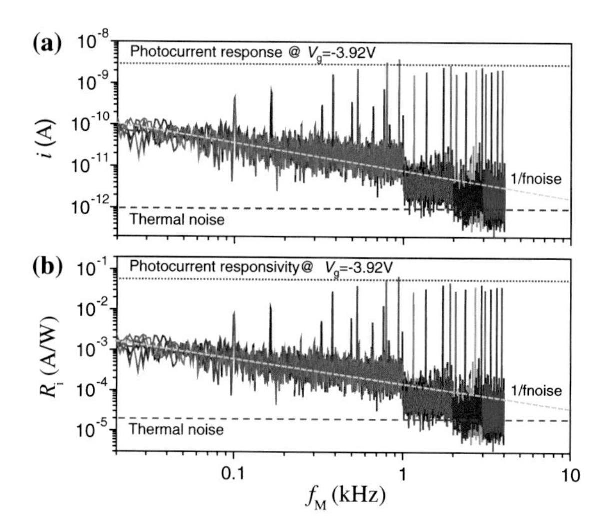

Fig. 3.29 **a** The photocurrent signal in Ampere and **b** the current responsivity as a function of the signal frequency. Both are calculated from Fig. 3.28

room-temperature bolometric detectors which are slow in speed, filed-effect terahertz detectors can offer a high speed. Kachorovskii and Shur have theoretically evaluated the maximum response speed for field-effect detectors [28]. The maximum speed can be formulated as

$$f_{\max} = \begin{cases} \frac{\mu V_{\text{geff}}}{2\pi L^2} & V_{\text{geff}} > 0,\ eV_{\text{geff}} \gg T \\ \frac{\mu \eta k_B T}{2\pi e L^2} & V_{\text{geff}} < 0,\ e\left|V_{\text{geff}}\right| \gg T, \end{cases} \tag{3.41}$$

where, V_{geff} the effective gate voltage, L is the gate length, e is the electron charge, $\eta \sim 1$ is the ideality factor, k_B the Boltzmann constant, and T is the temperature. Near $V_g = 0\,\text{V}, f_{\max}$ can be interpolated as

$$f_{\max} = \left(\frac{\mu\eta kT}{2\pi e L^2}\right)\left[1 + \exp\left(\frac{-eV_{\text{geff}}}{\eta k_B T}\right)\right] \ln\left[1 + \left(\frac{eV_{\text{geff}}}{\eta k_B T}\right)\right]. \tag{3.42}$$

The calculated maximum response frequency as a function of the effective gate voltage is shown in Fig. 3.30. For gate lengths of 200 nm, $f_{\max}$ is estimated to be about 65 GHz, 30 GHz, and 10 GHz for GaAs, GaN, and Si at the device threshold, respectively. Due to the strong dependence on the channel length ($1/L^2$), $f_{\max}$ is increased dramatically as the gate length becomes smaller. For example, a silicon FET with a gate length of 50 nm would allow detection at a speed higher than 20 GHz. The dashed line in Fig. 3.30 corresponds to the velocity saturation with $f_{\max} = \nu_s/2\pi L$ for $\nu_s = 1 \times 10^7$ cm/s.

An AlGaN/GaN self-mixing detector is compared with those of a silicon bolometer and a pyroelectric detector, as shown in Fig. 3.31. Both the self-mixing detector and the pyroelectric detector are operated at room temperature while the silicon bolometer is operated at 4.2 K. The response speed of the self-mixing detector is

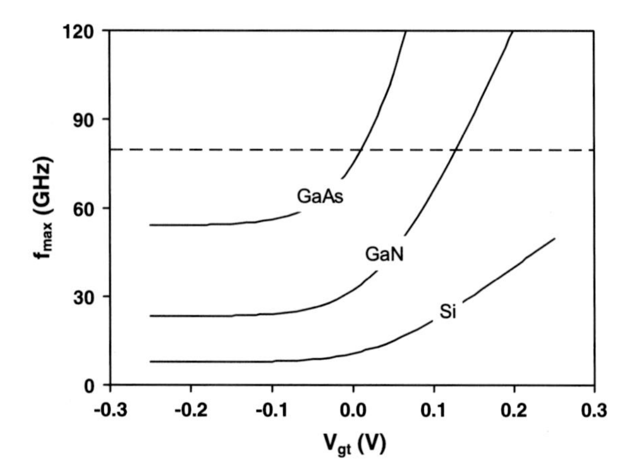

Fig. 3.30 The maximum response speed as a function of the effective gate voltage for different semiconductors at room temperature with a gate length of $L = 200\,\text{nm}$. Typical values of the mobility were used: $\mu = 3{,}500\,\text{cm}^2/\text{Vs}$ for GaAs, $\mu = 1{,}500\,\text{cm}^2/\text{Vs}$ for GaN and $\mu = 500\,\text{cm}^2/\text{Vs}$ for Si. The *dashed line* corresponds to the upper speed limit determined by the saturation velocity [28]

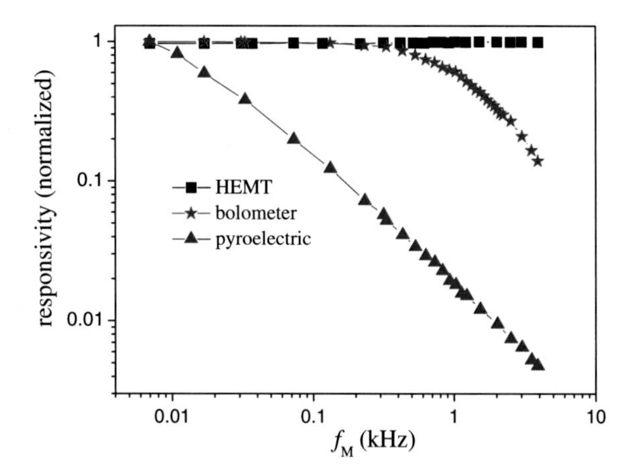

Fig. 3.31 The normalized detector signals as a function of the modulation frequency for the AlGaN/GaN self-mixing detector at room temperature, the silicon bolometer at 4.2 K and the pyroelectric detector at room temperature

much greater than the other two detectors. A slight roll-of in the current responsivity can be seen in Fig. 3.31 when the frequency gets higher than 1 kHz. The maximum modulation frequency about 4 kHz is limited by the electromechanical optical chopper. The frequency at the -3 dB roll-off is 600 Hz for the silicon bolometer and 30 Hz for the pyroelectric detector, respectively. Since, we do not have high-speed modulator for the BWO terahertz source, the actual maximum response speed cannot be measured directly.

In our case, it is the bandwidth of the current/voltage preamplifier for reading out the terahertz photocurrent/photovoltage that limits the detection speed. This fact is confirmed by measuring the different response speeds by setting the detector in voltage mode and current mode. As shown in Fig. 3.32, the normalized terahertz

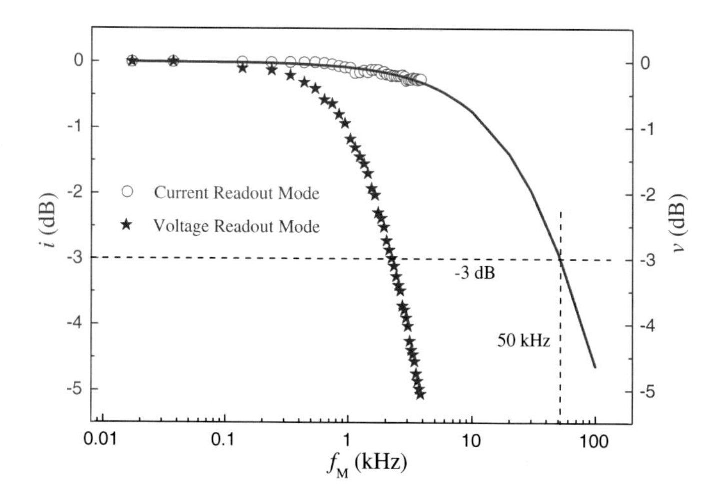

Fig. 3.32 Comparison of the normalized terahertz responses in the current mode and the voltage mode as a function of the modulation frequency

responses as a function of the modulation frequency are compared for the voltage mode and the current mode. The response bandwidth in the voltage mode is only 1.2 kHz while the bandwidth in the current mode reaches 50 kHz which is the bandwidth of the current preamplifier. In the voltage mode, although the bandwidth of the voltage preamplifier is about 1 MHz, the *RC* time constant of the detector connected to the voltage preamplifier by a coaxial cable becomes the dominant limiting factor for the response speed. The detector resistance is much smaller than the input impedance of the voltage preamplifier. The capacitance comes mainly from the coaxial line and the parasitic capacitance on the detector chip. The *RC* time constant in the current mode is not a concern since the input impedance of the current preamplifier is only a few Ohms. This additional parasitic effect of the readout circuit is further confirmed by measuring the terahertz response with different modulation frequencies. As shown in Fig. 3.33, no significant change is found in the current mode at different modulation frequencies and the maximum responsivity is fixed at $V_g = -3.9$ V. Very differently, the voltage responsivity depends strongly on the modulation frequency and the detector impedance. When the gate voltage approaches the pinch-off voltage, the impedance increases abruptly and the detector response is strongly suppressed by increasing the modulation frequency.

When a detector with high speed up to radio frequency (RF) is required, a voltage readout mode is preferred since RF low-noise amplifiers (LNAs) are widely available. However, the input impedance of such RF LNAs is usually matched to 50 Ω. Hence, the detector impedance needs to be matched with the input impedance of a RF LNA.

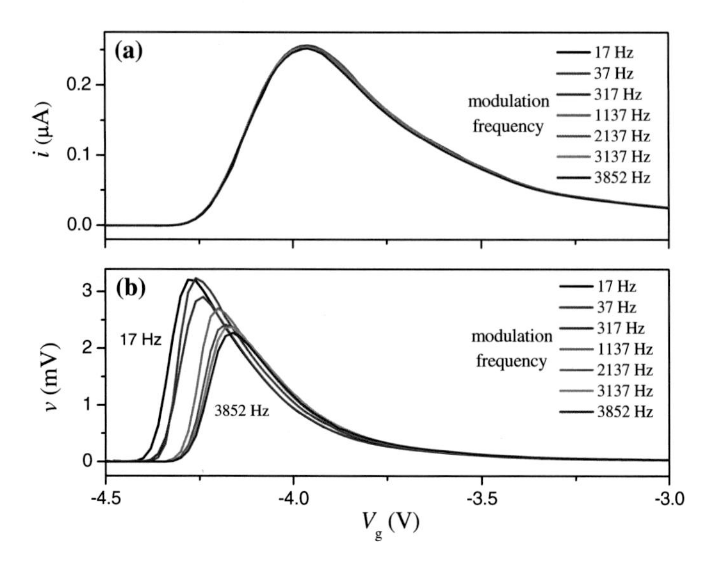

Fig. 3.33 **a** The current response and **b** the voltage response as a function of the gate voltage at different modulation frequencies

3.3.5 *Spectral Response*

For a specific terahertz application, the detector response is usually optimized in a given range of the spectrum. Our field-effect detector is designed for sensitive detection of the terahertz radiation generated by the BWO source which is in a range of 800–1000 GHz. The frequency response of the detector is measured by sweeping the output frequency of the BWO source, as shown in Fig. 3.34. Two frequency ranges are observed in 800–820 GHz and 870–920 GHz. Both are narrower than the simulated spectral range of the antenna. The discrepancy comes from the interference of the terahertz wave in the substrate of the detector which has a thickness of 416 μm and a refractive index of 3.4. Two simulated reflectance curves are shown in Fig. 3.34 by assuming the reflectance at the top interface (air-GaN) is 45 % and that at the bottom interface is 45 or 90 %. The observed ranges of the terahertz response fall into the frequency bands in which the reflectance is enhanced by interference. Such interference effect from the substrate could be eliminated by integrating the detector chip on a hyper-hemispherical silicon lens.

3.3.6 *Polarization-Dependent Response*

Terahertz responsivity of a field-effect detector is proportional to the mixing factor which in turn is defined by a terahertz antenna. The terahertz antenna we designed is composed of three quarter-wave dipole antennas. Hence, the detector response

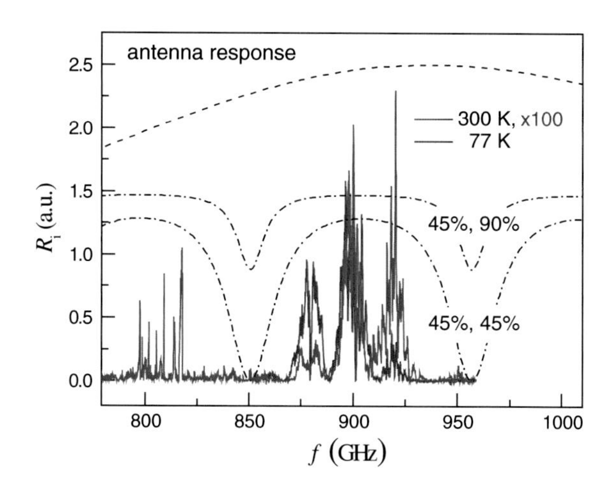

Fig. 3.34 The measured responsivity (*solid lines*). The simulated antenna response (*dashed line*) and the substrate interference (*dash-dotted lines*). Reprinted with permission from Ref. [5], copyright 2012, American Institute of Physics

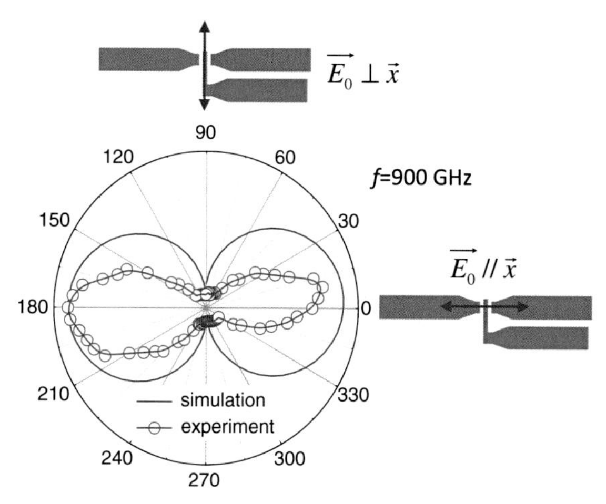

Fig. 3.35 The terahertz response at 900 GHz as a function of the angle between the polarization direction of the terahertz wave and the axial direction of the dipole antenna. The insert are the layout of the antenna. Reprinted with permission from Ref. [6], copyright 2011, American Institute of Physics

should be polarization dependent. The polarization-dependent terahertz response of our detector is shown in Fig. 3.35. The open circles are experiment data and the solid line is the theoretical dependence $\propto \cos^2\theta$, where orientation θ is defined as 0° or 180° when the polarization of the terahertz electric field is in parallel with the dipole antennas. The maximum terahertz response is observed when the polarization is along the length of the antennas as shown in Fig. 3.35. Such a polarization-dependent behavior is in agreement with the antenna design. M. Sakowicz pointed out the bonding pads and wires may become additional parts of the antenna and may affect the dependence on polarization [29]. Since two of the antenna blocks in our detector are floating and disconnected from the drain and the source electrodes, the bonding pads for the source and the drain have very little effect on the terahertz response unless they are placed in the near-field zone of the antennas. The only antenna block (g-antenna) is still connected to the bonding pad for the gate. The influence from the gate bonding pad can be minimized by using a terahertz low-pass filter as the electrical interconnect for the g-antenna and the corresponding bonding pad, as have been discussed in Chap. 2.

3.4 Detector Optimization

As we have shown in the detector model in Chap. 2, the responsivity is determined by the mixing factor and the field-effect factor which can be optimized by the antenna design, the Schottky gate and the 2DEG material. In this section, we emphasize the design rules for detector optimization by summarizing various field-effect detectors we experimented. Since the mobility of the AlGaN/GaN 2DEG has researched an optimal level and it becomes very difficult to be further improved, we focus on the design of terahertz antenna and its integration with the Schottky gate. A current responsivity of 71 mA/W, a voltage responsivity of 3.6 kV/W and an *NEP* of 40 $\mathrm{pW}/\sqrt{\mathrm{Hz}}$ are obtained through the development of five different types of detectors.

The five different types of the detectors are show in Fig. 3.36. The insets are zoom-in view of the central active regions where the arrangement of the Schottky gate and the antennas are illustrated. Asymmetric design is conducive to the asymmetric terahertz distribution in the electron channel which is essential for the enhancement of responsivity. The main characteristics of these five different detectors are described in the list below.

1. **Detector-A**, the first generation. As shown in Fig. 3.36a, square bonding pads are used as patch antennas. The size of the antenna is not optimized for the target terahertz frequency band around 900 GHz.
2. **Detector-B**, the second generation. Two dipole antennas are connected to the source and the gate to induce asymmetrical terahertz field in the gated electron channel, as shown in Fig. 3.36b.
3. **Detector-C**, the third generation. Three dipole blocks connected directly to the drain, the source, and the gate form an asymmetric antenna, as shown in Fig. 3.36c. Each antenna block is connected to the corresponding bonding pad via a single narrow and straight gold strip on the chip. The two ohmic contacts for the source and the drain are merged into the corresponding antenna block.
4. **Detector-D**, the fourth generation. To eliminate the influence from the bonding pads, three dipole blocks are connected to the source/drain/gate electrodes by using low-pass filters, as shown in Fig. 3.36d. Similar to Detector-C, the two ohmic contacts for the source and the drain are merged into the corresponding antenna block. Different to the other detectors which have a gate length of 2 μm, the gate length is 700 nm and the gap between the source antenna and the drain antenna is 3 μm.
5. **Detector-E**, the fifth generation. To further improve the antenna efficiency, the antenna blocks on top of the source mesa and the drain mesa are isolated from

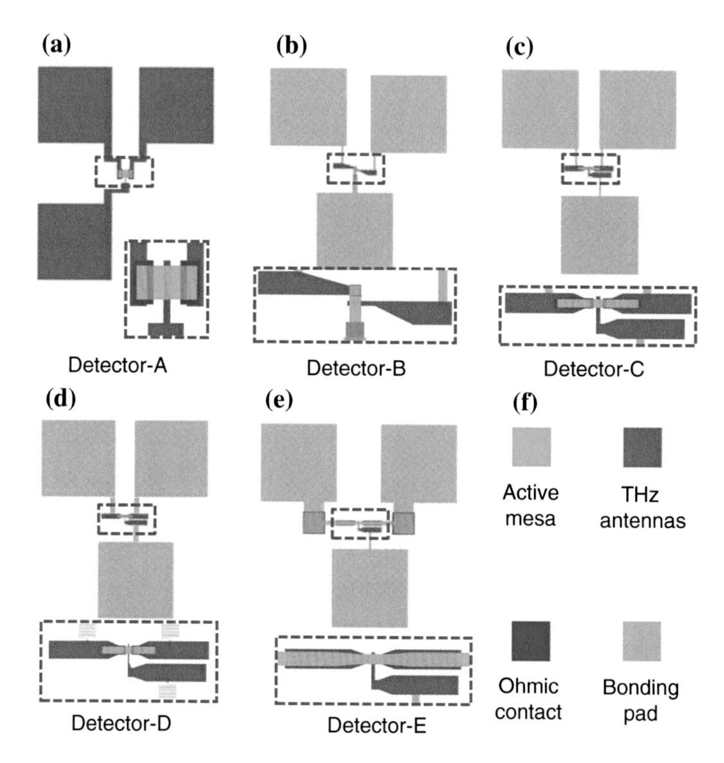

Fig. 3.36 Five generations of the field-effect detectors we studied. The insets are zoom-in view of the central active region where the arrangement of the Schottky gate and the antennas are illustrated. **a** Detector-A the first generation. **b** Detector-B the second generation. **c** Detector-C the third generation. **d** Detector-D the fourth generation. **e** Detector-E the fifth generation. **f** Colorized notations for the different detector components

the bonding pads, i.e., they are floating, as shown in Fig. 3.36e. The two ohmic contacts are set far away from the antennas. The gate antenna block is connected to the bonding pad via a narrow straight gold strip.

As shown in Fig. 3.37, the conductance, the field-effect factor and the responsivity for five different detectors are measured as a function of the gate voltage. For the sake of clarity the curves are shifted in the horizontal direction. The source–drain conductance measured without source–drain bias is shown in Fig. 3.37a. Detector-A has the maximal conductance since it is measured at 77 K. Due to the distance between the source/drain ohmic contacts is only 3 μm, the conductance of the detector-D is bigger than the other detectors at 300 K. On the contrary, the distance of the source/drain ohmic contacts is more than 130 μm for the detector-E, the conductance is smallest. The detector-D has a highest field-effect factor ($\mathrm{d}G/\mathrm{d}V_g$) at 300 K as shown in Fig. 3.37b except the detector-A measured at 77 K. The measured terahertz responsivity as a function of the gate voltage is shown in Fig. 3.37c. The responsivity is effectively improved through the five generations of detector design. The fifth

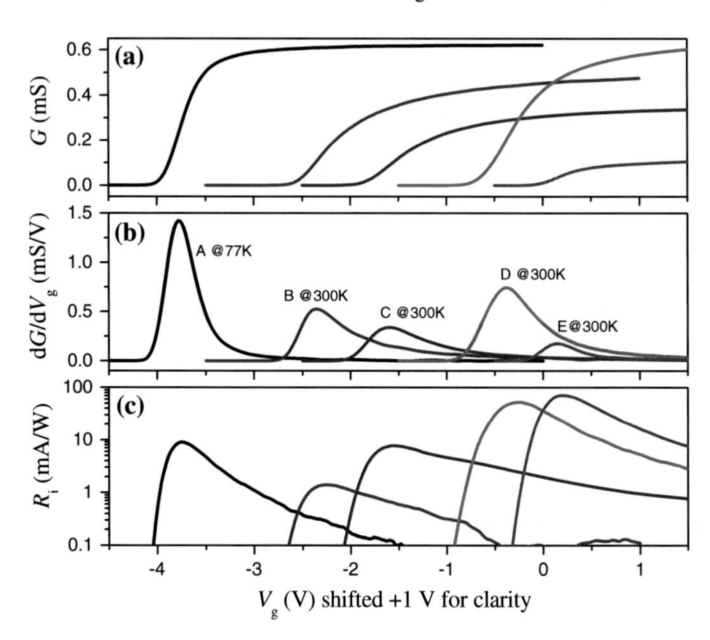

Fig. 3.37 Comparison of the different detectors with zero source–drain bias. **a** The measured source–drain conductance. **b** The derivative of the source–drain conductance. **c** The terahertz responsivity

Table 3.4 List of the main parameters for the five different detectors

No. (unit)	Antennas	Mesa length	W	L	$\frac{L}{W}\frac{dG}{dV_g}$	R_i or R_v	NEP
	–	(μm)	(μm)	(μm)	(mS/V)	(mA/W or V/W)	(nW/$\sqrt{\text{Hz}}$)
A	Patch antenna [8]	8	10	2	0.28 (77K)	9/100 (77K)	10 (77K)
B	Dipole antenna	11	5	2	0.21	1.4/30	10
C	Dipole antenna [5]	5	4	2	0.175	7.8/200	0.5
D	Filtered antenna [6]	3	4	0.7	0.13	53/1,000	0.2
E	Floating antenna [5]	130	4	2	0.09	71/3,600	0.04

generation, i.e., detector-E yields a current responsivity of 71 mA/W and a voltage responsivity of 3.6 kV/W.

The structural parameters, the electrical properties and the terahertz responsivity of these five detectors are listed in Table 3.4 for comparison. The responsivity of the detector-A is only 9 mA/W at 77 K since the patch antennas are based on the simple bonding pads and there is no intentional design of the feed line or the ground plane. The corresponding *NEP* is bigger than $10\,\text{nW}/\sqrt{\text{Hz}}$ at 77 K. Nevertheless, the detector-A is the very starting point of this thesis. Although the detector-B has an intentionally designed asymmetric antenna, the responsivity is only 1.4 mA/W at 300 K and the *NEP* is about $10\,\text{nW}/\sqrt{\text{Hz}}$, both are comparable with the detector-A. The antenna design with only two dipole blocks for the detector-B is not yet effective. For detector-C when three dipole blocks are incorporated into an asymmetric antenna both the horizontal and the perpendicular terahertz electric field are induced and enhanced in the electron channel. The responsivity is increased up to 7.8 mA/W and 200 V/W and the *NEP* is reduced to $0.5\,\text{nW}/\sqrt{\text{Hz}}$. By further reducing the gate length and the gap between the antennas in the detector-D, the current and voltage responsivity are increased to 53 mA/W and 1 kV/W, respectively. Although the antennas in detector-D are isolated from the bonding pads by low-pass filters, the terahertz near-field is affected by the ohmic contacts which are made at the end of the antennas. The rapid thermal annealing of the ohmic contacts induces uncontrolled or unreproducible morphology of the antenna end. In detector-E, the ohmic contacts are separated from the antennas and the antennas except the gate antenna are floating. In this way, the core structure in the central active region can be made in high quality in terms of the control of the antenna shape, the morphology, and the reproducibility. Although the distance between the ohmic contacts are much larger than the other detectors and hence the field-effect factor $\mathrm{d}G/\mathrm{d}V_{\mathrm{g}}$ is the smallest, the responsivity and the *NEP* are greatly improved to 3.6 kV/W and $40\,\text{pW}/\sqrt{\text{Hz}}$, respectively.

The above comparison among the five detectors with different antenna designs confirms that the mixing factor and hence the detector responsivity can be enhanced by optimizing the antenna design. Antennas other than the above discussed can be utilized for this purpose. The detector sensitivity not only depends on the responsivity but also depends on the noise level. Both responsivity and the noise level are related to the dimension of the electron channel and the quality of the 2DEG material including the electron mobility. It is highly expected that the responsivity can be further enhanced and detectors with $NEP < 1\,\text{pW}/\sqrt{\text{Hz}}$ can be made by optimizing the antennas together with the field-effect channel.

3.5 Summary

Details on the development and characterization of self-mixing terahertz detectors based on AlGaN/GaN 2DEG are presented in this chapter. The quasi-static model developed in Chap. 2 is experimentally verified by characterization. Two unique techniques are applied. One of them is the asymmetric antenna with three dipole blocks which induces strongly localized and asymmetrically distributed self-mixing factor ($\dot{\xi}_x\dot{\xi}_z \cos\phi$) in the electron channel. The other one is the manipulation of the localized field-effect factor ($\mathrm{d}n_x/\mathrm{d}V_{\mathrm{geff}}$) in the channel by tuning both the source–drain bias and the gate voltage. The source–drain bias is applied either on the source side or the drain side so that the spatial distributions of the electron density in these two cases are in a mirror symmetry. Such a control of the symmetry in the electron density distribution relative to the fixed symmetry of the self-mixing factor allows us to manipulate the polarity of the photocurrent and hence give direct evidences to verify the quasi-static detector model.

Optimization of the detectors is demonstrated mainly by searching more effective terahertz antennas. Five different detectors with specific antenna designs are fabricated, characterized, and compared to each other. The asymmetric antenna with three dipole blocks is confirmed to be the most efficient antenna among the five candidates. The effects of the bonding pads and the terahertz filters on the antenna efficiency are studied. Although an *NEP* down to $40\,\mathrm{pW}/\sqrt{\mathrm{Hz}}$ and a voltage responsivity of 3.6 kV/W at 900 GHz have been achieved, its does not mean the asymmetric antenna with three dipole blocks is the ultimate design. Based on the quasi-static detector model, the responsivity can be further improved by optimizing the antenna together with the field-effect channel including the gate.

Also it has to be mentioned that there is an obvious discrepancy found between the simulated and the experiment terahertz responses when the field-effect factor in the region where the self-mixing factor is the highest is suppressed. Further investigation on the model and more deliberately designed detector structures are required to uncover the origin of this ‘abnormal’ phenomenon. In Chap. 4, a detector with symmetric antenna and nanometer-sized gates is fabricated and characterized in a similar way to reveal the possible resonant excitation and detection of terahertz plasmons. The quasi-static detector model is further verified and resonant plasmon detection is observed together with the nonresonant self-mixing.

References

1. Ambacher, O.: Growth and applications of group-III-nitrides. J. Phys. D: Appl. Phys. **31**(20), 2653 (1998)
2. Nakamura, S., Senoh, M., Iwasa, N., Nagahama, S.: High-brightness InGaN blue, green and yellow light-emitting diodes with quantum well structures. Jpn. J. Appl. Phys. **34**, L797 (1995)
3. Nakamura, S., Makai, T., Sench, M.: High-brightness InGaN/AlGaN double-heterostructure blue-green-light-emitting diodes. J. Appl. Phys. **76**, 8189 (1994)
4. Puurunen, R.L.: Surface chemistry of atomic layer deposition: a case study for the trimethylaluminum/water process. J. Appl. Phys. **97**, 121301 (2005)
5. Sun, J.D., Sun, Y.F., Wu, D.M., Cai, Y., Qin, H., Zhang, B.S.: High-responsivity, low-noise, room-temperature, self-mixing terahertz detector realized using floating antennas on a GaN-based field-effect transistor. Appl. Phys. Lett. **100**, 013506 (2012)
6. Sun, Y.F., Sun, J.D., Zhou, Y., Tan, R.B., Zeng, C.H., Xue, W., Qin, H., Zhang, B.S., Wu, D.M.: Room temperature GaN/AlGaN self-mixing terahertz detector enhanced by resonant antennas. Appl. Phys. Lett. **98**, 252103 (2011)
7. Sun, J.D., Sun, Y.F., Zhou, Y., Zhang, Z.P., Lin, W.K., Zeng, C.H., Wu, D.M., Zhang, B.S., Qin, H., Li, L.L., Xu, W.: Enhancement of terahertz coupling efficiency by improved antenna design in GaN/AlGaN HEMT detectors. AIP Conf. Proc. **1399**, 893 (2011)

8. Zhou, Y., Sun, J.D., Sun, Y.F., Zhang, Z.P., Lin, W.K., Lou, H.X., Zeng, C.H., Lu, M., Cai, Y., Wu, D.M., Lou, S.T., Qin, H., Zhang, B.S.: Characterization of a room temperature terahertz detector based on a GaN/AlGaN HEMT. J. Semicond. **32**(4), 064005 (2011)
9. Tauk, R., Teppe, F., Boubanga, S., Coquillat, D., Knap, W., Meziani, Y.M., Gallon, C., Boeuf, F., Skotnicki, T., Fenouillet-Beranger, C., Maude, D.K., Rumyantseva, S., Shur, M.S.: Plasma wave detection of terahertz radiation by silicon field effects transistors: responsivity and noise equivalent power. Appl. Phys. Lett. **89**, 253511 (2006)
10. Knap, W., Dyakonov, M., Coquillat, D., Teppe, F., Dyakonova, N., Sakowski, J., Karpierz, K., Sakowicz, M., Valusis, G., Seliuta, D., Kasalynas, I., El Fatimy, A.: Field effect transistor for terahertz detection: physics and first imaging applications. J. Infrared Millim. Terahz. Waves **30**(12), 1319–1337 (2009)
11. Knap, W., Teppe, F., Meziani, Y., Dyakonova, N., Lusakowski, J., Buf, F., Skotnicki, T., Maude, D., Rumyantsev, S., Shur, M.S.: Plasma wave detection of sub-terahertz and terahertz radiation by silicon field-effect transistors. Appl. Phys. Lett. **85**, 675 (2004)
12. Lisauskas, A., Pfeiffer, U., Öjefors, E., Bolìvar, P.H., Glaab, D., Roskos, H.G.: Rational design of high-responsivity detectors of terahertz radiation based on distributed self-mixing in silicon field-effect transistors. J. Appl. Phys. **105**, 114511 (2009)
13. Knap, W., Rumyantsev, S., Lu, J., Shur, M., Saylor, C., Brunel, L.: Resonant detection of subterahertz radiation by plasma waves in a submicron field-effect transistor. Appl. Phys. Lett. **80**, 3433 (2002)
14. El Fatimy, A., Teppe, F., Dyakonova, N., Knap, W., Seliuta, D., Valuis, G., Shchepetov, A., Roelens, Y., Bollaert, S., Cappy, A., Rumyantsev, S.: Resonant and voltage-tunable terahertz detection in InGaAs/InP nanometer transistors. Appl. Phys. Lett. **89**, 131926 (2006)
15. Popov, V.V., Polischuk, O.V., Knap, W., El Fatimy, A.: Broadening of the plasmon resonance due to plasmon-plasmon intermode scattering in terahertz high-electron-mobility transistors. Appl. Phys. Lett. **93**, 263503 (2008)
16. Lü, J.Q., Shur, M.S.: Terahertz detection by high-electron-mobility transistor: enhancement by drain bias. Appl. Phys. Lett. **78**, 2587 (2001)
17. Veksler, D., Teppe, F., Dmitriev, A.P., Yu, V., Kachorovskii, Knap, W.: Detection of terahertz radiation in gated two-dimensional structures governed by dc current. Phys. Rev. B **73**, 125328 (2006)
18. Sun, J.D., Qin, H., Lewis, R.A., Sun, Y.F., Zhang, X.Y., Cai, Y., Wu, D.M., Zhang, B.S.: Probing and modelling the localized self-mixing in a GaN/AlGaN field-effect terahertz detector. Appl. Phys. Lett. **100**, 173513 (2012)
19. Tauk, R., Teppe, F., Boubanga, S., Coquillat, D., Knap, W.: Plasma wave detection of terahertz radiation by silicon field effects transistors: responsivity and noise equivalent power. Appl. Phys. Lett. **89**, 253511 (2006)
20. Öjefors, E., Lisauskas, A., Glaab, D., Roskos, H.G., Pfeiffer, U.R.: Terahertz imaging detectors in CMOS technology. J. Infrared Millim. Terahz. Waves **30**(12), 1269–1280 (2009)
21. Wang, B., Hellums, J.R., Sodini, C.G.: Thermal noise modeling for analog integrated circuits. IEEE J. Solid-State Circuits **29**, 833 (1994)
22. Hesler, J.L., Crowe, T.W.: Responsivity and noise measurements of zero-zias Schottky ziode zetectors. In: Proceeding of the 18th International Symposium on Space Terahertz Technology, vol. 18, pp. 89 (2007)
23. Pyroelectric detectors by gentec-eo. http://www.spectrumdetector.com/. Accessed 17 Jun 2014
24. Golay cells by MICROTECH instuments, Inc. http://www.mtinstruments.com/thzdetectors/index.htm. Accessed 12 May 2014
25. Boppel, S., Lisauskas, A., Krozer, V., Roskos, H.G.: Performance and performance variations of sub-1 THz detectors fabricated with 0.15 μm CMOS foundry process. Electron. Lett. **47**(11), 661–662 (2011)
26. Öjefors, E., Baktash, N., Zhao, Y., Hadi, R.A., Sherry, H., Pfeiffer, U.R.: Terahertz imaging detectors in a 65-nm CMOS SOI technology. In: 2010 Proceedings of 36th European Solid-State Circuits Conference, vol. 36, pp. 486–489 (2010)

27. Schuster, F., Coquillat, D., Videlier, H., Sakowicz, M., Teppe, F., Dussopt, L., Giffard, B., Skotnicki, T., Knap, W.: Broadband terahertz imaging with highly sensitive silicon CMOS detectors. Opt. Express **19**(8), 7827–7832 (2011)
28. Kachorovskii, V.Y., Shur, M.S.: Field effect transistor as ultrafast detector of modulated terahertz radiation. Solid State Electron. **52**(2), 182–185 (2008)
29. Sakowicz, M., sakowski, J., Karpierz, K., Grynberg, M., Knap, W., Gwarek, W.: Polarization sensitive detection of 100 GHz radiation by high mobility field-effect transistors. J. Appl. Phys. **104**, 024519 (2008)

Chapter 4
Realization of Resonant Plasmon Excitation and Detection

Abstract The effect of symmetries in the terahertz field distribution and the field-effect channel on terahertz photocurrent is examined and compared to the quasi-static field-effect detector model. Resonant excitation of cavity plasmon modes and nonresonant self-mixing of terahertz waves are demonstrated in an AlGaN/GaN two-dimensional electron gas with symmetrically designed nanogates, antennas, and filters. We found that the self-mixing signal can be effectively suppressed by the symmetric design and the resonant response benefits from the residual asymmetry. The findings further confirm the quasi-static field-effect detector model. The findings also suggest that a single detector may provide both a high sensitivity from the self-mixing mechanism and a good spectral resolution from the resonant response by optimizing the degree of geometrical and/or electronic symmetries.

4.1 Introduction

In Sect. 3.4, the nonresonant self-mixing of terahertz wave is realized in five different types of antenna-coupled field-effect electron channels based on AlGaN/GaN 2DEG. The quasi-static detector model and the effectiveness of the asymmetric antenna design are verified by probing the localized self-mixing photocurrent. Based on the fact that a high sensitivity of about 40 $\mathrm{pW}/\sqrt{\mathrm{Hz}}$ has been demonstrated at 900 GHz, it is foreseeable that the nonresonant self-mixing mechanism could be implemented in high-sensitivity and high-speed terahertz detectors for room temperature applications.

According to the hydrodynamic theories developed by M.I. Dyakonov, M.S. Shur, and others, resonant plasmon detection in field-effect electron channel may provide an even higher sensitivity [1–10]. On the other hand, the quasi-static self-mixing detector model suggests that nonresonant detection occurs in any field-effect detector as long as an asymmetric distribution of the horizontal and perpendicular terahertz fields are induced in the field-effect channel. Experimental results reported so far show that the resonant response is much weaker than the nonresonant self-mixing response [3–5]. Thus, it is somehow difficult to distinguish the resonant response from the nonresonant response because the former is usually submerged into the

J. Sun, *Field-effect Self-mixing Terahertz Detectors*, Springer Theses,
DOI 10.1007/978-3-662-48681-8_4

latter. However, unlike the nonresonant self-mixing detection with a maximum at a certain gate voltage at which the field-effect factor $\mathrm{d}n/\mathrm{d}V_\mathrm{g}$ also reaches its maximum, the resonant response is maximized when the plasmon frequency (f_p) tuned by the gate voltage equals the terahertz frequency (f). At different terahertz frequencies, the optimal gate voltage for the maximal resonant response is different according to the following dispersion relation for a gated plasmon cavity [1, 2]

$$f_\mathrm{p} = \frac{1}{4L_\mathrm{eff}}\sqrt{\frac{eV_\mathrm{geff}}{m^*}}, \tag{4.1}$$

where L_eff is the effective length of the plasmon cavity, e and m^* are the electron charge and effective mass, respectively. The amplitude of the resonant response is limited by the quality factor ($Q = \omega_p\tau_\mathrm{p}$, where ω_p is the plasmon's angular frequency and τ_p is the plasmon's relaxation time) of the plasmon resonance. In the resonant case ($Q_\mathrm{p} = \omega_\mathrm{p}\tau_\mathrm{p} \gg 1$), a plasma wave is excited resonantly at a proper gate voltage and induces a unidirectional photocurrent which has a Lorentzian line shape [2–5]

$$i_\mathrm{R} = I_\mathrm{T}\frac{f_\mathrm{p}^2}{(f - f_\mathrm{p})^2 + (1/4\pi\tau_\mathrm{p})^2}, \tag{4.2}$$

where prefactor I_T is a complex function of the incident terahertz power, channel conductance, the geometry, and the boundary conditions. Understanding the physics of plasmon damping would allow us to control the plasmon cavity with proper boundary conditions and to develop high-performance plasmon-based terahertz devices.

As we have shown in Chap. 3, in the nonresonant case ($\omega_\mathrm{p}\tau_\mathrm{p} \ll 1$), the photoresponse comes from the self-mixing of the terahertz electric field in the electron channel [11–17]. The short-circuit self-mixing photocurrent can be rewritten as [16]

$$i_\mathrm{M} = \frac{E_0^2}{4}\bar{z}\int_0^L \frac{\mathrm{d}G}{\mathrm{d}V_\mathrm{geff}}\dot{\xi}_x\dot{\xi}_z \cos\phi \,\mathrm{d}x, \tag{4.3}$$

where E_0 is the free-space electric field induced by the incident THz wave, $G = G(x)$ is the local conductance of the gated channel, $\bar{z}$ is the effective distance between the gate and the channel, and L is the length of the channel. $V_\mathrm{geff} = V_\mathrm{g} - V_\mathrm{th} - V_x$ is effective gate voltage, V_g, V_th and V_x are the applied voltage, the threshold voltage, and the local channel potential, respectively. $\dot{\xi}_x$, $\dot{\xi}_z$, and ϕ are the horizontal and perpendicular terahertz field enhancement factors, and the phase difference between the induced fields, respectively. In an antenna coupled field-effect detector, the total photocurrent can be expressed as

$$i_\mathrm{T} = i_\mathrm{M} + i_\mathrm{R}. \tag{4.4}$$

Both nonresonant self-mixing and resonant plasmon responses rely on the asymmetries in the terahertz field distribution or the plasmon cavity's boundary conditions.

For resonant plasmon detection, the source/drain ohmic contacts usually serve as the boundaries for the plasmon cavity and also as part of the terahertz antennas. The asymmetric boundary conditions for plasmon wave can also be achieved by biasing the device with a direct current source [6–8]. For nonresonant self-mixing detection, as shown in Chap. 3, the asymmetric condition is reached by either using asymmetric antennas or biasing the electron channel with a proper current [11, 15]. In this chapter, we double check the effect of symmetry on the terahertz photocurrent in an AlGaN/GaN field-effect detector with symmetrically arranged nanogates, antennas, and low-pass filters. The three nano gates form a tunable plasmon cavity with nearly symmetric boundaries. The metallic source/drain contacts are set 1.75 μm away from the plasmon cavity to maintain the degree of symmetry. In this detector, both self-mixing response and resonant response are observed. By suppressing the nonresonant self-mixing response, the resonant response is clearly resolved at 77 K. The terahertz photocurrent as a function of the gate voltage and the source-drain bias is mapped for the verification of the symmetric design.

4.2 Detector with a Symmetric Antenna

A scanning-electron-microscope graph of the detector is shown in Fig. 4.1. Three nanogates are defined as G_1, G_2, and G_3. A plasmon cavity is formed between gates G_2 and G_3. Gate G_1 applied with voltage V_g is used to control the carrier density and thus tune the plasmon frequency. Antenna A_1 and A_2 are designed to have a resonant frequency at 900 GHz and are connected to gate G_1 and G_2/G_3, respectively. The meander-shaped low pass filters are used to connect the antennas with the corresponding wire-bonding pads [18]. The gate length is 150, 100, and

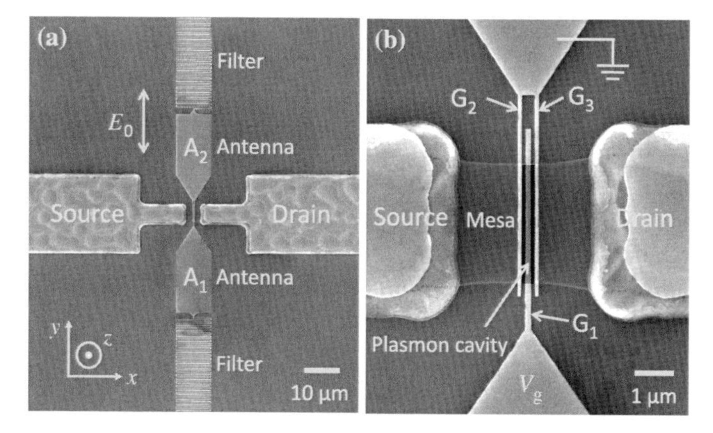

Fig. 4.1 **a** Scanning-electron-microscope graph of the field-effect terahertz detector. **b** Zoom-in view of the central active region including the plasmon cavity, the nanogates, and the 2DEG channel. Reprinted with permission of Ref. [19], copyright 2015, American Institute of Physics

100 nm for gate G_1, G_2, and G_3, respectively. The distance between the inner edges of gate G_2 and gate G_3 is about 330 nm. The width of the electron channel (mesa) is $W = 3$ μm and the distance between the source and drain contacts is 3.5 μm. At 77 K, the electron mobility and the electron density are $\mu = 1.58 \times 10^4$ cm^2/Vs and $n_s = 1.06 \times 10^{13}$ cm^{-2}, respectively. The effective electron mass is $m^* = 0.2\,m$, where m is the free electron mass. The radiation from a BWO terahertz source is modulated by a mechanical chopper at frequency $f_M = 317$ Hz. The short-circuit photocurrent is measured using a current preamplifier and a lock-in amplifier. The photocurrent is maximized when the polarization of the linearly polarized terahertz wave becomes parallel to the antennas along direction y.

4.3 Antenna Simulation

As shown in Fig. 4.2, we present numerical calculations of the terahertz electric field and the phase distributions at 900 GHz by a finite-difference time-domain (FDTD) method. To obtain the fine field distribution in the gated 2DEG, the grid constant is set

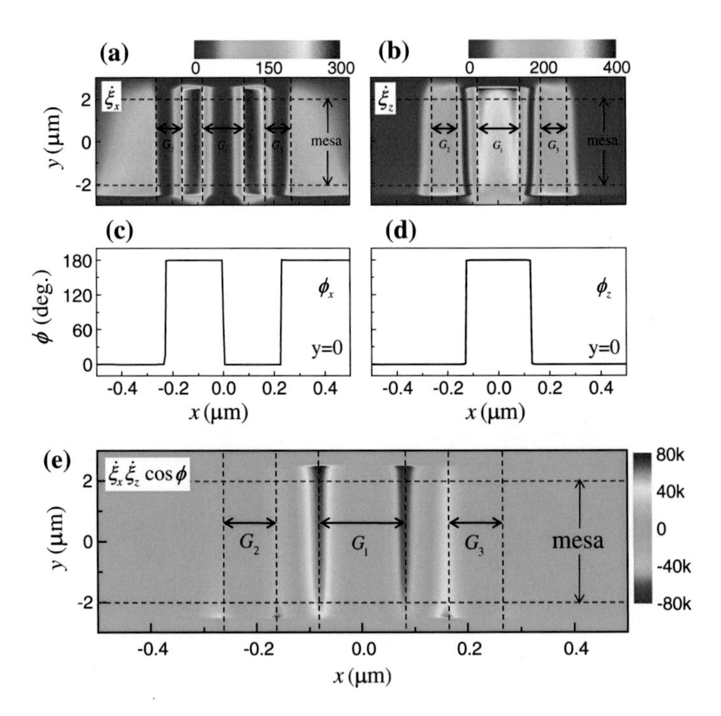

Fig. 4.2 Spatial distribution of the terahertz electric field and the phase from an FDTD simulation at 900 GHz. **a** Field enhancement factor $\dot{\xi}_x$. **b** Field enhancement factor $\dot{\xi}_z$. **c**, **d** The phases ϕ_x and ϕ_z at $y = 0$. **e** Mixing factor $\dot{\xi}_x \dot{\xi}_z \cos\phi$. Reprinted with permission of Ref. [19], copyright 2015, American Institute of Physics

as 5 and 20 nm in direction x and direction y in FDTD simulations, respectively. The field enhancement factors for the horizontal and the perpendicular fields are shown in Fig. 4.2a, b, respectively. The horizontal field is concentrated in the extended areas between gates G_1, G_2, and G_3, while the perpendicular field is mainly distributed under gate G_1. The simulation confirms that, unlike the asymmetric detector shown in Chap. 3 [16], the source and drain sides have the same electric field distribution. The horizontal field changes its phase by π at the center of the three gates, while the perpendicular field changes its phase at the center of the gaps between the gates, as shown in Fig. 4.2c and d. In Fig. 4.2e, the simulated mixing factor ($\dot{\xi}_x \dot{\xi}_z \cos\phi$) is plotted to reveal the spatial distribution. Strong mixing symmetrically occurs only at the left and right edges of gate G_1. However, the mixing at the source side generates a photocurrent in an opposite direction to that induced at the drain side. This inversion of polarity comes from an phase flip. Since the device structure is symmetric along the electron channel, the self-mixing response is expected to be suppressed effectively according to Eq. 4.3.

4.4 Resonant Plasmon Detection

As shown in Fig. 4.3a, the source–drain conductance G and its derivative $\mathrm{d}G/\mathrm{d}V_\mathrm{g}$ as a function of the gate voltage are characterized at 77 K without a source–drain bias. The derivative is maximized at −4.00 V. Figure 4.3b shows the photocurrent versus V_g at 850 GHz (□), 861 GHz (○), 907 GHz (△), and 940 GHz (▽). For the

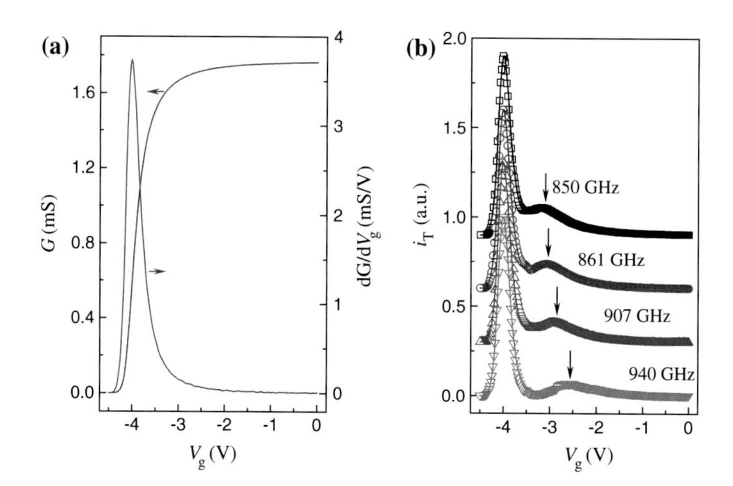

Fig. 4.3 **a** Channel conductance and its derivative as a function of the gate voltage. **b** Terahertz response at different terahertz frequencies (□: 850 GHz, ○: 861 GHz, △: 907 GHz, ▽: 940 GHz). The *solid curves* are calculated responses based on Eq. 4.4. The *arrows mark* the locations of plasmon resonances. Reprinted with permission of Ref. [19], copyright 2015, American Institute of Physics

sake of clarity the curves are shifted in the vertical scale. Two peaks are observed with one of them fixed at −4.00 V and the other at a lower gate voltage around −3 V, depending on the terahertz frequency. The frequency-independent response at the same location where the derivative of the conductance is maximal comes from the nonresonant self-mixing as described by Eq. 4.3. The frequency-dependent response is induced by the excitation of cavity plasmons and can be well described by Eq. 4.2. The higher the terahertz frequency, the higher the peak gate voltage (i.e., the higher the electron density). When we tune the terahertz frequency from 850 to 940 GHz, the position of the resonant response shifts from −3.12 to −2.60 V. The solid curves in Fig. 4.3b are calculated photocurrents based on Eq. 4.4 and agree well with the experiments. By normalizing the total photoresponse and substracting the self-mxing response from the measured total photocurrent, the net resonant responses at different terahertz frequencies are shown in Fig. 4.4a. The resonant frequency as a function of the gate voltage fits well with Eq. 4.1, as shown in Fig. 4.4b. In the fitting, the size of the plasmon cavity is chosen to be $L_{\mathrm{eff}} = 330$ nm, i.e., the gap size between gate G_2 and G_3. As a comparison, the dashed line shown in Fig. 4.4b represents the gate-voltage-tuned plasmon frequency with a cavity size of 150 nm, i.e., the length of gate G_1. This implies that plasmons are excited and confined in the 2DEG enclosed by G_2 and G_3 instead of in the 2DEG right beneath the control gate G_1 [3–5].

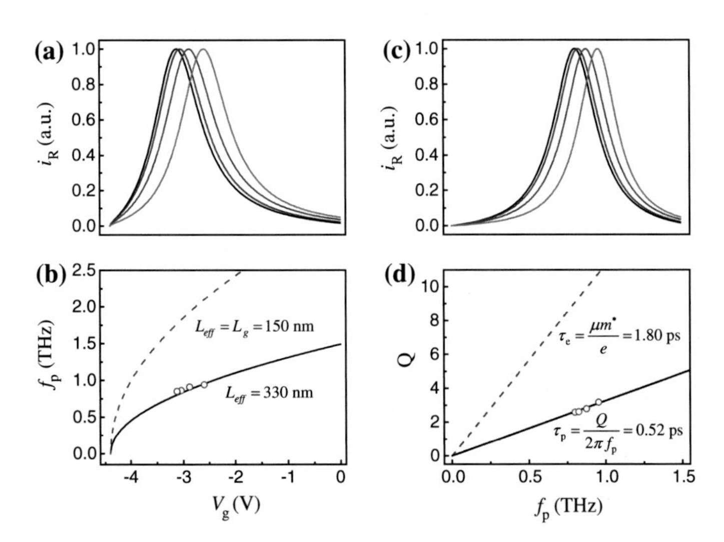

Fig. 4.4 **a** Resonant photocurrent as a function of the gate voltage. **b** Resonant plasmon frequency as a function of the gate voltage. The *solid curve* and the *dashed curve* correspond to an effective cavity length of $L_{\mathrm{eff}} = 330$ nm and $L_{\mathrm{eff}} = 150$ nm, respectively. **c** Resonant photocurrent as a function of the plasmon frequency. **d** Quality factor of the plasmon resonance extracted from (**c**) as a function of the plasmon frequency. The *solid curve* and the *dashed curve* correspond to a plasmon relaxation time of $\tau_p = 0.52$ ps and a transport relaxation time of $\tau_e = 1.80$ ps, respectively. Reprinted with permission of Ref. [19], copyright 2015, American Institute of Physics

As we can see in Fig. 4.3b, the resonant response is much weaker than the nonresonant self-mixing response. The current responsivity is estimated to be 540 and 59 mA/W for the nonresonant self-mixing response and the resonant response, respectively. The corresponding noise-equivalent powers are 3 $\mathrm{pW}/\sqrt{\mathrm{Hz}}$ and 45 $\mathrm{pW}/\sqrt{\mathrm{Hz}}$. Compared to the detectors with a 2-μm gate at room temperature shown in Chap. 3 [15], this nano-gated detector exhibits an unexpectedly low responsivity in the self-mixing detection mode at 77 K. It is the symmetric antenna design that strongly suppresses the self-mixing response. The amplitude of the resonant response relies on how plasmons are effectively excited and how effectively the charge oscillations are converted into a direct current. The former can be characterized by the plasmon quality factor $Q_\mathrm{p} = 2\pi f_\mathrm{p}\tau_\mathrm{p}$. The latter is a complex function of the boundary conditions of the plasmon cavity and it is not yet specifically known. By substituting Eq. 4.1 into Eq. 4.2, we obtain the resonant response as a function of the plasmon frequency (Fig. 4.4c) and the plasmon quality factors can be extracted as shown in Fig. 4.4d. We found a spectral resolution of better than 10 GHz, which can be achieved by tuning the gate voltage. By fitting the experimentally extracted quality factors, we found the effective plasmon lifetime is $\tau_\mathrm{p} = 0.52$ ps which is about one-third of the electron's relaxation time $\tau_\mathrm{e} = \mu m^*/e = 1.80$ ps obtained from the Hall mobility at 77 K. This indicates that damping of plasmon due to strong electron–electron and electron–plasmon interactions takes place in the cavity [5].

4.5 The Effect of Symmetry

To further verify the symmetric design, we use the same technique as that introduced in Chap. 3 [16]. The degree of symmetry in the photocurrent mapped in a 2D color-scale plot as a function of V_ds and V_g at 907 GHz is shown in Fig. 4.5a. A strong self-mixing photocurrent is produced near −4.00 V and is marked with a dashed line. When a large positive bias is applied at the drain (source) side, the 2DEG under the gate at the drain (source) side is depleted and the locally induced negative (positive) photocurrent is suppressed effectively, as schematically shown in the insets of Fig. 4.5b. Along the dashed line in Fig. 4.5a, the measured photocurrent comes from the locally induced positive (negative) photocurrent at the source (drain) side of the gated 2DEG channel. The corresponding current responsivity increases from 0.54 A/W to 1.72 A/W as bias V_D changes from 0 to 0.5 V. The absolute photocurrent as a function of V_ds is extracted along the dashed line and is plotted in Fig. 4.5b. The maximum photocurrent with the bias applied at the drain and at the source is 83 nA and 57 nA, respectively. The expected location of the minimal photocurrent is shifted from 0 V to $V_\mathrm{S} = 0.1$ V. At $V_\mathrm{ds} = 0$ V, the photocurrent is about 26 nA, which is about the difference between the observed maximal responses when the bias is applied at the drain and at the source. This confirms that a residual asymmetry exists in our symmetrically designed device. It is this asymmetry that allows us to bring up the resonant response with a suppressed self-mixing photoresponse as the background. Since the broadband nonresonant response is always expected to

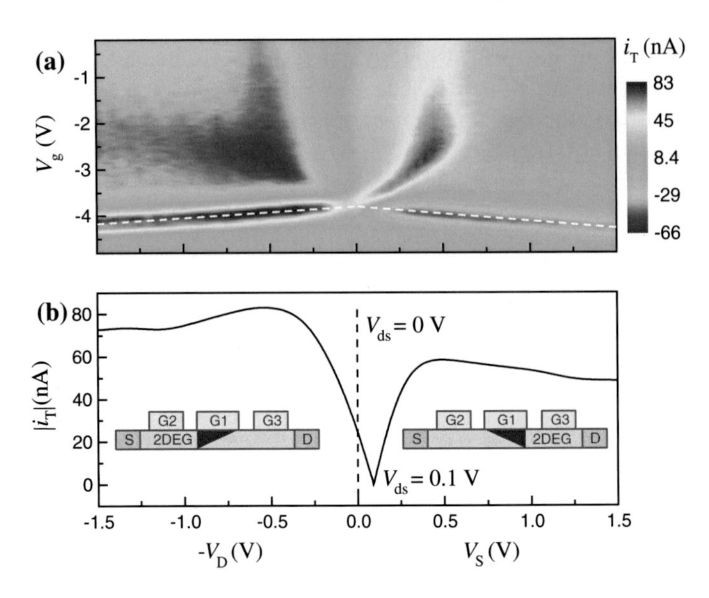

Fig. 4.5 **a** Examination of the degree of symmetry in the terahertz photocurrent mapped in a 2D color-scale plot as a function of V_{ds} and V_g at 907 GHz. **b** Absolute terahertz photocurrent as a function of V_{ds} extracted along the *dashed line* in (**a**). Reprinted with permission of Ref. [19], copyright 2015, American Institute of Physics

occur, it is difficult to separate the resonant response from the nonresonant response when both the gate voltage and the source–drain bias are applied. However, in our symmetrically designed detector we can set the detector at a proper operation point so that the residual asymmetry in self-mixing response can be electrically cancelled and minimize the self-mixing response, in our case $V_S = 0.1$ V. On the other hand, by operating the detector with a large source–drain bias, the detector allows for terahertz detection with an enhanced sensitivity. Thus, a symmetrically designed detector could provide both high sensitivity and a spectral resolution.

4.6 Summary

In this chapter, we have demonstrated and analyzed both the resonant and nonresonant (self-mixing) terahertz photoresponses in an AlGaN/GaN field-effect detector with symmetrically arranged nanogates, antennas, and filters. Experiments confirm that the self-mixing response can be effectively suppressed by the symmetric design. The observed self-mixing response at zero bias suggests that a certain degree of asymmetry exists in the terahertz field distribution and/or in the field-effect electron channel. The observed resonant response agrees well with the quarter-wavelength plasmon mode defined by the nanogates. Although this resonant photoresponse may be induced by the residual asymmetry in our symmetrically designed device, more in-depth research is required for further engineering the boundary conditions for cavity plasmon modes. The broadband self-mixing mechanism and the narrowband resonant detection integrated in one detector may provide both high sensitivity and spectral resolution.

References

1. Dyakonov, M.I., Shur, M.S.: Shallow water analogy for a ballistic field effect transistor: new mechanism of plasma wave generation by dc current. Phys. Rev. Lett. **71**, 2465 (1993)
2. Dyakonov, M., Shur, M.S.: Detection, mixing, and frequency multiplication of terahertz radiation by two-dimensional electronic fluid. IEEE Trans. Electron Devices **43**(3), 380–387 (1996)
3. Knap, W., Rumyantsev, S., Lu, J., Shur, M., Saylor, C., Brunel, L.: Resonant detection of subterahertz radiation by plasma waves in a submicron field-effect transistor. Appl. Phys. Lett. **80**, 3433 (2002)
4. El Fatimy, A., Teppe, F., Dyakonova, N., Knap, W., Seliuta, D., Valuis, G., Shchepetov, A., Roelens, Y., Bollaert, S., Cappy, A., Rumyantsev, S.: Resonant and voltage-tunable terahertz detection in InGaAs/InP nanometer transistors. Appl. Phys. Lett. **89**, 131926 (2006)
5. Popov, V.V., Polischuk, O.V., Knap, W., El Fatimy, A.: Broadening of the plasmon resonance due to plasmon-plasmon intermode scattering in terahertz high-electron-mobility transistors. Appl. Phys. Lett. **93**, 263503 (2008)
6. Boubanga-Tombet, S., Teppe, F., Coquillat, D., Nadar, S., Dyakonova, N., Videlier, H., Knap, W., Shchepetov, A., Gardes, C., Roelens, Y., Bollaert, S., Seliuta, D., Vadoklis, R., Valueis, G.: Current driven resonant plasma wave detection of terahertz radiation: toward the Dyakonov-Shur instability. Appl. Phys. Lett. **92**, 212101 (2008)
7. Peralta, X.G., Allen, S.J., Wanke, M.C., Harff, N.E., Simmons, J.A., Lilly, M.P., Reno, J.L., Burke, P.J., Eisenstein, J.P.: Terahertz photoconductivity and plasmon modes in double-quantum-well field-effect transistors. Appl. Phys. Lett. **81**, 1627 (2002)
8. Dyer, G.C., Aizin, G.R., Allen, S.J., Grine, A.D., Bethke, D., Reno, J.L., Shaner, E.A.: Induced transparency by coupling of Tamm and defect states in tunable terahertz plasmonic crystals. Nature Photon. **7**, 925–930 (2013)
9. Tanigawa, T., Onishi, T., Takigawa, S., Otsuji, T.: Enhanced responsivity in a novel AlGaN/GaN plasmon-resonant terahertz detector using gate-dipole antenna with parasitic elements. In: The 68th Device Research Conference, vol. 68, pp. 167–168 (2010)
10. Kim, S., Zimmerman, J.D., Focardi, P., Gossard, A.C., Wu, D.H., Sherwin, M.S.: Room temperature terahertz detection based on bulk plasmons in antenna-coupled GaAs field effect transistors. Appl. Phys. Lett. **92**, 253508 (2008)
11. Knap, W., Deng, Y., Rumyantsev, S., Shur, M.S.: Resonant detection of subterahertz and terahertz radiation by plasma waves in submicron field-effect transistors. Appl. Phys. Lett. **81**, 4637 (2002)
12. Otsuji, T., Shur, M.: Terahertz plasmonics: good results and great expectations. IEEE Microw. Mag. **15**, 43 (2014)
13. Lisauskas, A., Pfeiffer, U., Öjefors, E., Bolìvar, P.H., Glaab, D., Roskos, H.G.: Rational design of high-responsivity detectors of terahertz radiation based on distributed self-mixing in silicon field-effect transistors. J. Appl. Phys. **105**, 114511 (2009)
14. Hadi, R.A., Sherry, H., Grzyb, J., Zhao, Y., Förster, W., Keller, H.M., Cathelinand, A., Kaiser, A., Pfeiffer, U.R.: A 1 k-pixel video camera for 0.7–1.1 terahertz imaging applications in 65 nm CMOS. IEEE J. Solid-State Circuits **47**, 2999 (2012)
15. Sun, J.D., Sun, Y.F., Wu, D.M., Cai, Y., Qin, H., Zhang, B.S.: High-responsivity, low-noise, room-temperature, self-mixing terahertz detector realized using floating antennas on a GaN-based field-effect transistor. Appl. Phys. Lett. **100**, 013506 (2012)

16. Sun, J.D., Qin, H., Lewis, R.A., Sun, Y.F., Zhang, X.Y., Cai, Y., Wu, D.M., Zhang, B.S.: Probing and modelling the localized self-mixing in a GaN/AlGaN field-effect terahertz detector. Appl. Phys. Lett. **100**, 173513 (2012)
17. Vicarelli, L., Vitiello, M.S., Coquillat, D., Lombardo, A., Ferrari, A.C., Knap, W., Polini, M., Pellegrini, V., Tredicucci, A.: Graphene field-effect transistors as room-temperature terahertz detectors. Nat. Mater. **11**, 865–871 (2012)
18. Sun, Y.F., Sun, J.D., Zhou, Y., Tan, R.B., Zeng, C.H., Xue, W., Qin, H., Zhang, B.S., Wu, D.M.: Room temperature GaN/AlGaN self-mixing terahertz detector enhanced by resonant antennas. Appl. Phys. Lett. **98**, 252103 (2011)
19. Sun, J.D., Qin, H., Lewis, R.A., Yang, X.X., Sun, Y.F., Zhang, Z.P., Li, X.X., Zhang, X.Y., Cai, Y., Wu, D.M., Zhang, B.S.: The effect of symmetry on resonant and nonresonant photoresponses in a field-effect terahertz detector. Appl. Phys. Lett. **106**, 031119 (2015)

Chapter 5
Scanning Near-Field Probe for Antenna Characterization

Abstract In the terahertz regime, the active region for a solid-state detector usually needs to be implemented accurately in the near-field region of an on-chip antenna. Mapping of the near-field strength could allow for rapid verification and optimization of new antenna/detector designs. Here, we report a proof-of-concept experiment in which the field mapping is realized by a scanning metallic probe and a fixed AlGaN/GaN field-effect transistor. Experiment results agree well with the electromagnetic-wave simulations. The results also suggest that a field-effect terahertz detector combined with a probe tip could serve as a high-sensitivity terahertz near-field sensor.

5.1 Introduction

In the terahertz portion of the electromagnetic spectrum, many sensing applications, such as security screening and near-field microscopy and spectroscopy, are being studied and developed [1]. Sensitive detectors are one of the key elements for such applications [2]. In various terahertz detectors, antennas are commonly applied to feed incident terahertz electromagnetic radiation into the active region of the detectors [3–9]. The efficiency of these antennas is crucial for high sensitivity and is largely determined by the near-field properties. The near-field distribution is usually obtained by performing finite-element analysis [10–13]. From the point of view of detector optimization/development, it would be beneficial to experimentally obtain the near-field distribution. In many terahertz near-field experiments/applications, the near-field electromagnetic wave is transferred by either a sharpened metallic probe tip or by a metallic pin-hole aperture into the far field and detected therein [14–17]. In terahertz time-domain spectroscopy, a photoconductive detector has been integrated on a probe tip and serves as a direct near-field terahertz detector [18]. In this chapter, we present experiment results on imaging the near-field response of an antenna-coupled field-effect terahertz detector by scanning the antennas using a sharpened metallic tip. In the experiment, the scanning metallic tip serves as a near-field coupler/agitator of the antennas and the integrated field-effect channel reads out the intensity of the terahertz wave. The experimental results are in good agreement

J. Sun, *Field-effect Self-mixing Terahertz Detectors*, Springer Theses,
DOI 10.1007/978-3-662-48681-8_5

with a finite-difference time-domain (FDTD) simulation. This method allows one to experimentally distinguish the most effective antenna block and provides direct guidance for the optimization of antennas and detectors.

5.2 Scanning Probe Setup

The experimental setup is schematically shown in Fig. 5.1a, where the terahertz radiation from a backward wave oscillator (BWO) is collected, collimated, and focused by a pair of off-axis parabolic mirrors (OAP#1 and OAP#2). Two optical photos of the setup are shown in Fig. B.2 in Appendix B. The terahertz frequency is set at $f_0 = 875$ GHz corresponding to a free-space wavelength of $\lambda_0 = 343$ μm. The schematic zoom-in view of the metallic probe scanning the detector surface is shown in Fig. 5.1b. The terahertz wave is polarized in direction x. The metallic probe is glued on a piezoelectric vibrator which is driven by a sinusoidal voltage with frequency of 123 Hz and peak-to-peak voltage of 5 V. The probe tip vibrates in direction x and the vibration amplitude is $\delta x_p \approx 1$ μm. Accurate positioning and raster scanning of the probe tip at a certain distance to the detector surface are realized by mounting the piezoelectric vibrator on a step-motorized XYZ stage. The probe is made of Tungsten and, as shown in Fig. 5.1c, has a diameter of 500 μm. The radius of the probe tip is sharpened to about 0.5 μm by electrochemical etching in NaOH solution.

The detector is similar to that studied in Chap. 3 [3]. A partial top view of the detector is shown in Fig. 5.2. The active electron channel is made up of an AlGaN/GaN 2DEG and the width in direction y is $W = 10$ μm. The antenna contains three blocks (**A**, **B**, and **C**). Each block is 45 μm long in direction x and maximally 14 μm wide in direction y. Only block **C** is directly connected to the external electronics via the electrode for applying the gate voltage. Blocks **A** and **B** are capacitively coupled to the 2DEG channel. The gate in the center of the detector controls the electron density

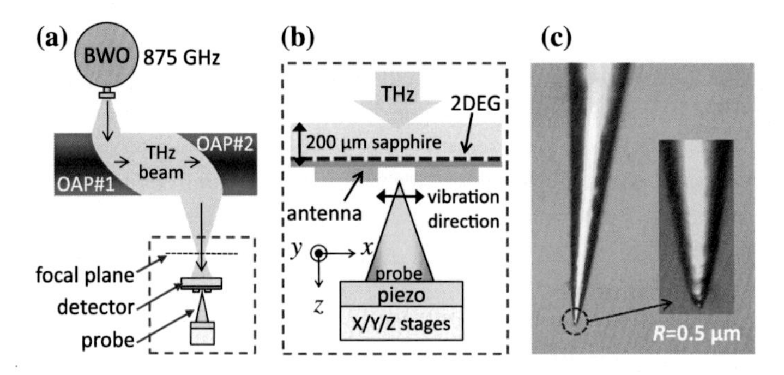

Fig. 5.1 **a** Schematics of the experiment setup. The continuous-wave terahertz source is tuned at 875 GHz. **b** Zoom-in view of the near-field probing scheme using a motorized vibrating probe. **c** Optical microscope images of the metallic probe and the tip. Reprinted with permission from Ref. [19], copyright 2015, Chinese Physical Society

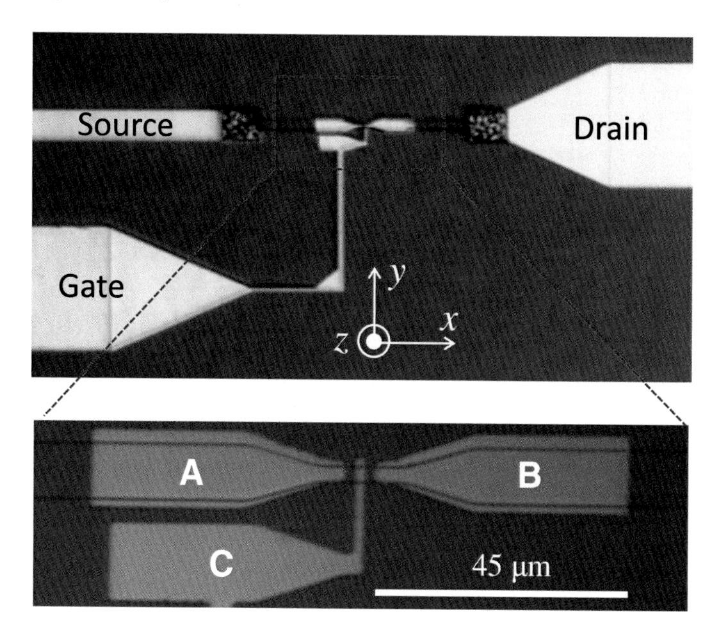

Fig. 5.2 Optical microscope image of the field-effect terahertz detector. The zoom-in box shows the central active region including three antenna blocks (**A**, **B**, **C**), the gate, and the 2DEG channel. Reprinted with permission from Ref. [19], copyright 2015, Chinese Physical Society

underneath and has a length of $L = 2$ μm in direction x. The ohmic contacts for the 2DEG channel are about 100 μm away from the gate. The sapphire substrate of the detector is transparent for the incident terahertz wave and is thinned to 200 μm. A high-resolution power microscope is used to monitor the probe tip and the terahertz detector. The net photocurrent induced by the vibrating probe is amplified by a current preamplifier, and then read out by a lock-in amplifier. By raster scanning the probe tip at distance z_p to the detector surface, the tip-induced photocurrent can be mapped as a function of the tip location (x_p, y_p).

The detector is first characterized by measuring the photocurrent without the probe tip upon terahertz irradiation of $f_0 = 875$ GHz. The photocurrent (i_T) as a function of the gate voltage (V_G) is shown in Fig. 5.3a. The peak photoresponse is located at $V_G = -3.55$ V. For the following experiments, the gate voltage is fixed at this optimal value. The profile of the terahertz beam, as shown in Fig. 5.3b, was obtained by raster scanning the detector in the focal plane of OAP#2 with a step size of 50 μm. The minimum beam width is about 1.4 mm and is about 15 times larger than the overall antenna dimension (≈90 μm) in direction x. In the following experiments, the detector is moved slightly away from the focal plane so that the whole antenna area is under a rather uniform irradiation, as schematically shown in Fig. 5.1a.

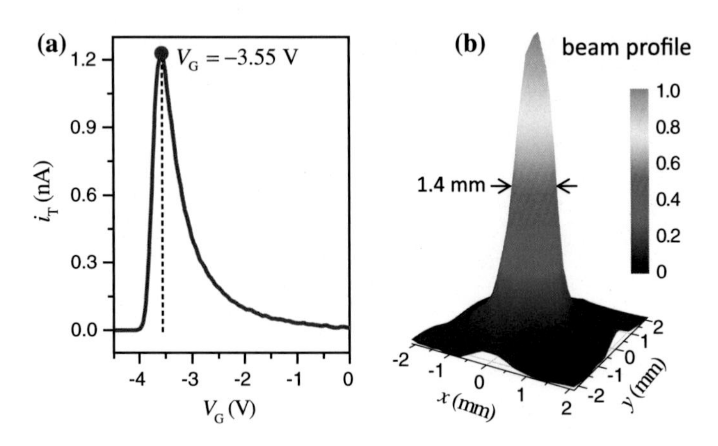

Fig. 5.3 **a** The photocurrent (i_T) tuned by the gate voltage (V_G). **b** The terahertz beam profile probed by the detector scanned in the focal plane of OAP#2. Reprinted with permission from Ref. [19], copyright 2015, Chinese Physical Society

5.3 Scanning Near-Field Photocurrent

A photocurrent image as shown in Fig. 5.4a is obtained by scanning the tip with $z_p = 0.5$ μm. The antenna profile is overlaid on the map for easy pattern recognition. Each antenna block shows a different terahertz response to the scanning tip. The

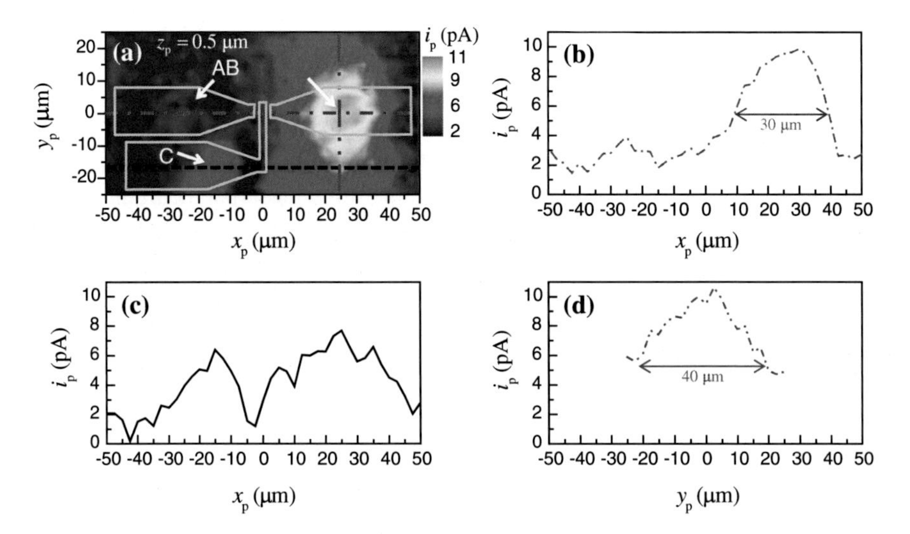

Fig. 5.4 **a** Map of the tip-induced photocurrent (i_p) by raster scanning the tip with a tip-detector distance of $z_p = 0.5$ μm. The profile of the antennas and the gate is overlaid on the map. **b** A line scan alone the *dash-dotted line* in (**a**), i.e., with $y_p = 0$ μm. **c** A line scan alone the *dashed line* in (**a**), i.e., with $y_p = -17.5$ μm. **d** A line scan alone the *dash-dot-dotted line* in (**a**), i.e., with $x_p = 25$ μm. Reprinted with permission from Ref. [19], copyright 2015, Chinese Physical Society

maximum response is observed at $(x_p, y_p) = (25\ \mu\text{m}, 0\ \mu\text{m})$, i.e., at the center of antenna block **B**. When the probe tip is located near the center of antenna block **C**, a weaker, but yet significant, photocurrent is observed. In comparison to antenna blocks **B** and **C**, the vibrating probe tip above antenna block **A** induces a much weaker photocurrent. The signal-to-noise ratio is only 2.7 corresponding to the noise floor of 1.5 pA. For clarity, two line scans along direction x and one scan along direction y were extracted from Fig. 5.4a. The corresponding locations of these scans are marked by the dash-dotted ($y_p = 0\ \mu\text{m}$), dashed ($y_p = -17.5\ \mu\text{m}$), and dash-dot-dotted line ($x_p = 25\ \mu\text{m}$) in Fig. 5.4a and plotted in Fig. 5.4b–d. The line scan at $y_p = 0\ \mu\text{m}$, crossing both antenna **A** and **B**, is shown in Fig. 5.4b. Two peaks at $x_p \approx \pm 25\ \mu\text{m}$ are clearly identified corresponding to the center points of antenna **A** and antenna **B**, respectively. The full-width-at-half-maximum (FWHM) of the peak originated from antenna **B** is about 30 μm, corresponding to a spatial resolution about $\lambda_0/11$, which is slightly smaller than the antenna width in direction y. The line scan at $y_p = -17.5\ \mu\text{m}$, crossing only antenna **C**, is shown in Fig. 5.4c. Two peaks with similar amplitude are shown clearly. The line scan at $x_p = 25\ \mu\text{m}$ crossing antenna **B** is shown in Fig. 5.4d. The FWHM in direction y is about 40 μm, slightly smaller than the antenna width in direction y and slightly larger than the FWHM in direction x.

With a tip-antenna distance of $z_p = 0.5\ \mu\text{m}$, different roles of three antenna blocks can be roughly identified as have been shown in Fig. 5.4a. The tip-induced photocurrent was further examined at larger tip-detector distances. As shown in Fig. 5.5a–d, four raster scans were obtained with $z_p = 1, 5, 15$, and 25 μm, respectively. We found that the smaller the tip-detector distance, the stronger the photocurrent is. When the tip-detector distance is greater than 15 μm, the detector becomes very insensitive to

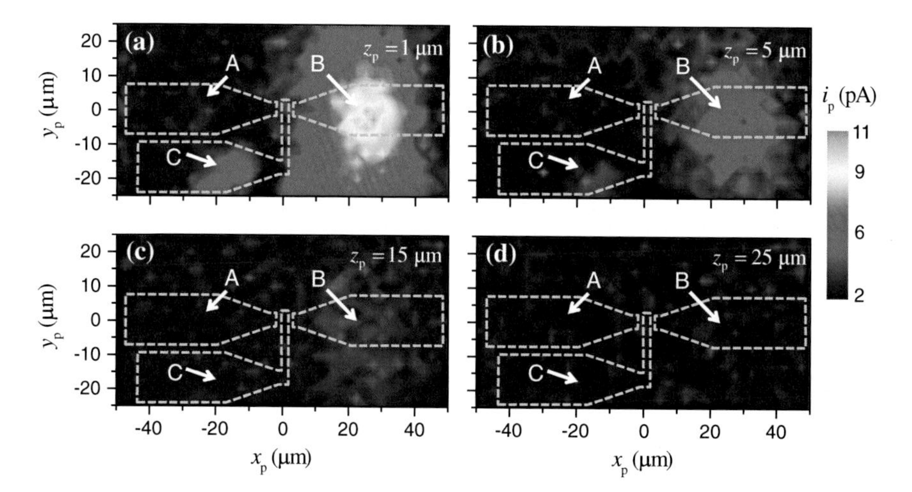

Fig. 5.5 Maps of the tip-induced photocurrent by raster scanning the tip at different tip-detector distances: **a** $z_p = 1\ \mu\text{m}$, **b** $z_p = 5\ \mu\text{m}$, **c** $z_p = 15\ \mu\text{m}$, and **d** $z_p = 25\ \mu\text{m}$. Reprinted with permission from Ref. [19], copyright 2015, Chinese Physical Society

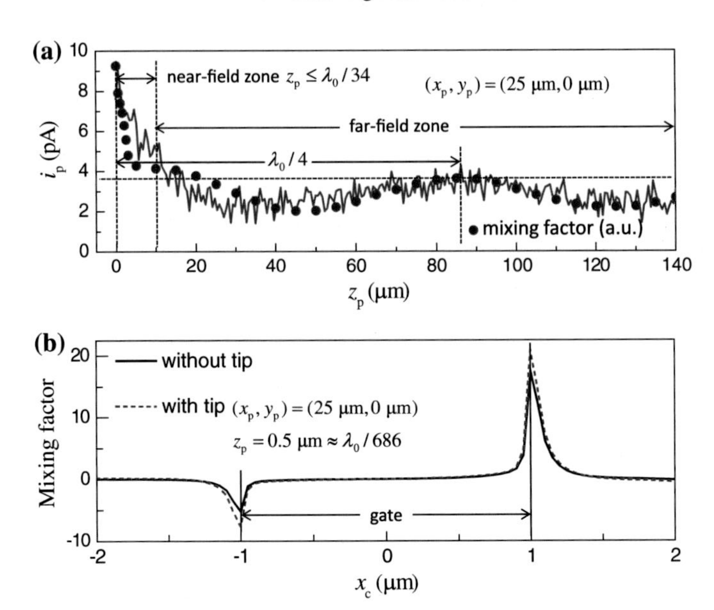

Fig. 5.6 **a** Tip-induced photocurrent as a function of the tip-antenna distance. The tip is pointed to the center of antenna block **B**, i.e., at $(x_p, x_y) = (25\ \mu m, 0\ \mu m)$. The solid data points in an arbitrary unit are simulated mixing factors at $x_c = 1\ \mu m$. **b** Simulated mixing factors in the 2DEG channel with and without the near-field probe tip. In both simulations, the probe-antenna distance is set at $z_p = 0.5\ \mu m$. Reprinted with permission from Ref. [19], copyright 2015, Chinese Physical Society

the probe tip. At a distance of 25 μm, the induced photocurrent is submerged in the background noise. There is an inactive region near the right-angle bend of antenna **C**, where a minimum photocurrent is observed. This feature clearly confirms that antenna **B** and antenna **C** have an opposite polarization.

The interaction between the tip and antenna block **B** was further examined by probing the photocurrent as a function of distance z_p when the tip is pointed to the center of antenna **B**. As shown in Fig. 5.6a, by moving the probe tip away from antenna block **B** to a distance about $10\ \mu m \approx \lambda/34$, the photocurrent decreases abruptly indicating a near-field interaction mechanism. On further increasing the distance, an oscillation in the photocurrent was observed. A minimum occurs at $z_p = 46\ \mu m \approx \lambda/7.5$ and a maximum occurs at $z_p = 86\ \mu m \approx \lambda/4$.

According to nonresonant self-mixing detection mechanism, the terahertz photocurrent is proportional to the integral of the local mixing factor ($\eta = \dot{\xi}_x \dot{\xi}_z \cos\phi$), the incident terahertz power flux (P_0), and the extent of electron density tuned by the gate voltage (dn/dV_G): $i_T = \alpha \int_0^L \eta\, dx$, where $\alpha = (e\mu W Z_0 P_0 \bar{z}/2L) dn/dV_G$, e is the elementary charge, μ is the electron mobility, n is the electron density of the 2DEG, Z_0 is the free-space impedance, and $\bar{z}$ is an effective distance between the antenna and the 2DEG. Factors $\dot{\xi}_x$ and $\dot{\xi}_z$ correspond to the terahertz field enhancement in directions x and z. Phase ϕ is the phase difference between the two field components. Vibrating the probe tip in direction x modulates the local terahertz electric fields and hence the mixing factor. Although the specific manner of how the tip alters the terahertz field is unknown, we can model the tip-induced photocurrent as

$$i_\mathrm{p} = \frac{\mathrm{d}i_\mathrm{T}}{\mathrm{d}x_\mathrm{p}} \times \delta x_\mathrm{p} = \delta x_\mathrm{p} \times \alpha \int_0^L \frac{\mathrm{d}\eta}{\mathrm{d}x_\mathrm{p}}\,\mathrm{d}x. \tag{5.1}$$

In order to obtain an insight into the mechanism of the tip-antenna interaction, FDTD simulations were performed to determine the differences in the mixing factor depending whether the tip is present or not, as shown in Fig. 5.6b. The simulation result suggests that mixing occurs predominately around $x_\mathrm{c} = \pm 1\ \mu\mathrm{m}$, i.e., near the edges of the gated 2DEG channel. Furthermore, the mixing factor at $x_\mathrm{c} = +1\ \mu\mathrm{m}$ is greater than that at $x_\mathrm{c} = -1\ \mu\mathrm{m}$. The simulation also suggests that the near-field tip coupled to antenna block **B** enhances the mixing factor by 46 and 19.6 % at $x_\mathrm{c} = -1\ \mu\mathrm{m}$ and at $x_\mathrm{c} = +1\ \mu\mathrm{m}$, respectively. This confirms that the antennas predominately determine the mixing factors and the probe tip induces only a perturbation. Since it requires a huge mesh to simulate the detector-tip configuration, it is not practical to make a full simulation of term $\mathrm{d}\eta/\mathrm{d}x_\mathrm{p}$. Nevertheless, simulations of the mixing factor as a function of the tip-antenna distance were performed. The simulated data points in an arbitrary unit are overlaid on the experimental tip-induced photocurrent as shown in Fig. 5.6a. Both the near-field response and the interference effect are recovered. However, there is a remarkable deviation between the simulation and the experimental data when $z_\mathrm{p} < 10\ \mu\mathrm{m}$, i.e., within the near-field zone. The deviation is most probably from the insufficient mesh number in the simulation since the near-field property is very sensitive to the distance and the shape of the tip.

From the raster scans shown in Figs. 5.4a and 5.5, we infer that block **B** is the most effective part of the terahertz antenna and the other two blocks are complementary. Our experiment also suggests that a terahertz near-field sensor maybe constructed by integrating a sharp metallic tip with antenna **B**. In this way, the tip serves as a scanning near-field antenna, antenna block **B** serves as a terahertz transmission line, and the gate-controlled field-effect channel functions as a sensitive detector. Such an integrated near-field sensor may provide a simple solution for terahertz microscope with high resolving power.

In current experiment, the tip vibrates in direction x and the vibration amplitude is much less than the antenna dimension. The antenna is less sensitive to the location of the tip than to the tip-antenna distance. According to the simulation shown in Fig. 5.6b, a tip vibrating in the z direction would allow for more straightforward probing of the near-field effect. For this reason, a modified probe system using a quartz tuning fork to excite the tip vibration in the z direction is under development. This system may allow for spontaneous imaging of the detector morphology in the atomic-force-microscope mode.

5.4 Summary

In this chapter, we have performed a near-field imaging experiment on an antenna-coupled field-effect terahertz detector. The combined usages of a scanning metallic probe and the field-effect terahertz detector allow us to image the active region of terahertz antennas. This method provides an alternative way for rapid verification of terahertz antenna design. Furthermore, an integrated scanning terahertz sensor may be developed for high-resolution terahertz microscopes.

References

1. Sizova, F., Rogalski, A.: THz detectors. Prog. Quantum Electron. **34**, 278 (2010)
2. Tonouchi, M.: Cutting-edge terahertz technology. Nat. Photon. **1**, 97–105 (2007)
3. Sun, J.D., Sun, Y.F., Wu, D.M., Cai, Y., Qin, H., Zhang, B.S.: High-responsivity, low-noise, room-temperature, self-mixing terahertz detector realized using floating antennas on a GaN-based field-effect transistor. Appl. Phys. Lett. **100**, 013506 (2012)
4. Sun, Y.F., Sun, J.D., Zhang, X.Y., Qin, H., Zhang, B.S., Wu, D.M.: Enhancement of terahertz coupling efficiency by improved antenna design in GaN/AlGaN high electron mobility transistor detectors. Chin. Phys. B **21**(10), 108504 (2012)
5. Tanigawa, T., Onishi, T., Takigawa, S., Otsuji, T.: Enhanced responsivity in a novel AlGaN / GaN plasmon-resonant terahertz detector using gate-dipole antenna with parasitic elements. In: The 68th Device Research Conference, vol. 68, pp. 167–168 (2010)
6. Lisauskas, A., Mundt, M., Seliuta, D., Minkevicius, L., Kasalynas, I., Valusis, G., Mittendorff, M., Winnerl, S., Krozer, V., Roskos, H.G.: CMOS integrated antenna-coupled field-effect transistors for the detection of radiation From 0.2 to 4.3 THz. IEEE Trans. Microwave Theory **60**, 3834 (2012)
7. Vicarelli, L., Vitiello, M.S., Coquillat, D., Lombardo, A., Ferrari, A.C., Knap, W., Polini, M., Pellegrini, V., Tredicucci, A.: Graphene field-effect transistors as room-temperature terahertz detectors. Nat. Mater. **11**, 865–871 (2012)
8. Dyer, G.C., Vinh, N.Q., Allen, S.J., Aizin, G.R., Mikalopas, J., Reno, J.L., Shaner, E.A.: A terahertz plasmon cavity detector. Appl. Phys. Lett. **97**, 193507 (2010)
9. Kim, S., Zimmerman, J.D., Focardi, P., Gossard, A.C., Wu, D.H., Sherwin, M.S.: Room temperature terahertz detection based on bulk plasmons in antenna-coupled GaAs field effect transistors. Appl. Phys. Lett. **92**, 253508 (2008)
10. Sun, J.D., Qin, H., Lewis, R.A., Sun, Y.F., Zhang, X.Y., Cai, Y., Wu, D.M., Zhang, B.S.: Probing and modelling the localized self-mixing in a GaN/AlGaN field-effect terahertz detector Appl. Phys. Lett. **100**, 173513 (2012)
11. Deibel, J.A., Escarra, M., Berndsen, N., Wang, K., Mittleman, D.M.: Finite-element method eimulations of guided wave phenomena at terahertz frequencies. Proc. IEEE **95**(8), 1624–1640 (2007)
12. Chen, H.T., Padilla, W.J., Zide, J.M., Gossard, A.C., Taylor, A.J., Averitt, R.D.: Active terahertz metamaterial devices. Nature **444**, 597–600 (2006)
13. Yang, K., David, G., Yook, J.G., Papapolymerou, I., Katehi, L.P., Whitaker, J.F.: Electrooptic mapping and finite-element modeling of the near-field pattern of a microstrip patch antenna. IEEE Trans. Microwave Theory **48**(2), 288–294 (2000)
14. Rosner, B.T., van der Weide, D.W.: High-frequency near-field microscopy. Rev. Sci. Instrum. **73**, 2505 (2002)
15. Chen, H.T., Kersting, R., Cho, G.C.: Terahertz imaging with nanometer resolution. Appl. Phys. Lett. **83**, 3009 (2003)

16. Mitrofanov, O., Bener, I., Harel, R., Wynn, J.D., Pfeiffer, L.N., West, K.W., Federici, J.: Terahertz near-field microscopy based on a collection mode detector. Appl. Phys. Lett. **77**, 3496 (2000)
17. Mitrofanov, O., Lee, M., Hsu, J.W., Pfeiffer, L.N., West, K.W., Wynn, J.D., Federici, J.F.: Terahertz pulse propagation through small apertures. Appl. Phys. Lett. **79**, 907 (2001)
18. Kawano, Y., Ishibashi, K.: An on-chip near-field terahertz probe and detector. Nat. Photon. **2**, 618–621 (2008)
19. Lü, L., Sun, J.D., Lewis, R.A., Sun, Y.F., Wu, D.M., Cai, Y., Qin, H.: Mapping an on-chip terahertz antenna by a scanning near-field probe and a fixed field-effect transistor. Chin. Phys. B. **24**(2), 028504 (2015)

Chapter 6
Applications

Abstract Nonresonant self-mixing terahertz detectors have a few advantages concerning the room-temperature operation, high sensitivity, wide response spectrum, high speed, and high signal-to-noise ratio over other slow bolometric detectors for room-temperature applications. Terahertz imaging based on single-pixel detectors and a 1×9 linear detector array are demonstrated with a high spatial resolution and a high signal-to-noise ratio. The single-pixel detector is successfully implemented in a Fourier transform spectrometer and a fast Fourier transform spectroscopy of a 900 GHz terahertz signal is demonstrated. Although the demonstrated self-mixing detectors are far from optimal in sensitivity, speed, stability, and reproducibility, this technology has a great potential to be further developed and optimized to bring many terahertz applications at room temperature into reality.

6.1 Single-Pixel Terahertz Imaging

Because of the unique nature of terahertz waves, terahertz imaging is finding more and more applications [1–9]. A variety of terahertz imaging system have been developed based on either pulsed, continuous wave (CW), or quasi-CW terahertz sources [10–23]. Most of the terahertz imaging systems are limited by either the low sensitivity or the low speed of the detector. Nonresonant self-mixing detectors have the potential for both high-speed and high-sensitivity detection at room temperature [24–29]. Also, self-mixing detectors can be made to sense a wide range (0.1–5 THz) of the terahertz spectrum [30, 31]. They can be applied to various types of coherent terahertz sources such as terahertz gas lasers, frequency multipliers, THz-TDS, Gunn-diode oscillators, backward-wave oscillators (BWOs), quantum cascade lasers (QCLs), and etc. [32–36]. On the other hand, large-scale focal plane arrays of self-mixing detectors may give rise to high-speed terahertz cameras [37, 38]. Here, we demonstrate raster-scan terahertz imaging and line-scan imaging using a single-pixel self-mixing detector and a 1×9 linear detector array, respectively. Both transmission-type and reflection-type raster-scan imaging tests are performed based on the single-pixel detector. Only transmission-type imaging is tested for the linear detector array. Both

J. Sun, *Field-effect Self-mixing Terahertz Detectors*, Springer Theses,
DOI 10.1007/978-3-662-48681-8_6

types of the imaging devices are designed for 900 GHz terahertz sources and the BWO is used as the illumination for imaging.

6.1.1 Terahertz Transmission Imaging

The setup for the raster-scan imaging in transmission mode is shown in Fig. 6.1. A photo setup is shown in Fig. B.3 in Appendix B. The terahertz wave is collected by two off-axis parabolic mirrors (OAPs) and focused onto the objects to be imaged. The terahertz wave penetrates through the objects and is then collected by another pair of OAP mirrors before it is focused onto the single-pixel self-mixing detector. The objects are concealed in a paper envelope. The raster scan of the envelope is realized by placing the envelop on a two-axis step-motorized stage. To achieve a high signal-to-noise ratio, the terahertz wave is modulated by a mechanical chopper and the detector signal is amplified by a current preamplifier and then read out by a lock-in amplifier with an integration time of 5 ms.

The first imaging test is to see the terahertz beam profiles at the focal point. As a comparison, both a commercial pyroelectric detector and an AlGaN/GaN self-mixing detector are evaluated using the above setup. As shown in Fig. 6.2a, the effective detection area of the pyroelectric detector is 2 mm $\times$ 2 mm marked by the dashed square. It is difficult to achieve high spatial resolution using this detector and the full-width-at-half-maximum (FWHM) of the profile seen is larger than 2 mm. In contrast, the pixel size of the self-mixing detector is about 20 μm $\times$ 97 μm (with antennas) and a much smaller beam profile is obtained as shown in Fig. 6.2b. Figure 6.2c, d displays the beam profiles in 3D. Since the dimension of the self-mixing detector is smaller than the terahertz wavelength (333 μm), the scanned image represents the actual beam profile at the focal point which has a FWHM of 500 μm. The spatial resolution of the imaging is limited by the beam size.

The second imaging test is to see through the envelop with different objects concealed within. As shown in Fig. 6.3a, the transmission terahertz image is compared

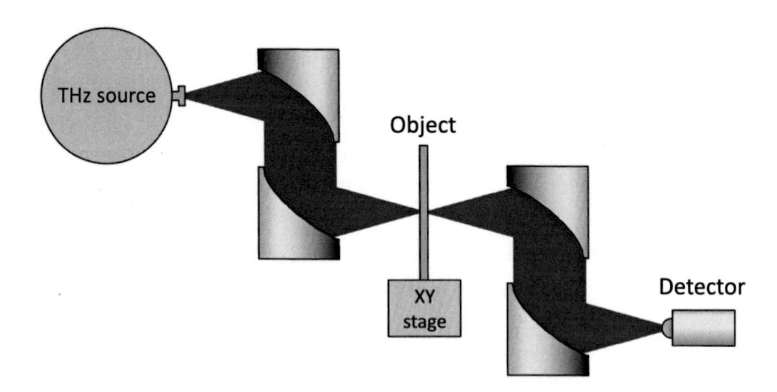

Fig. 6.1 Raster-scan imaging setup in transmission mode

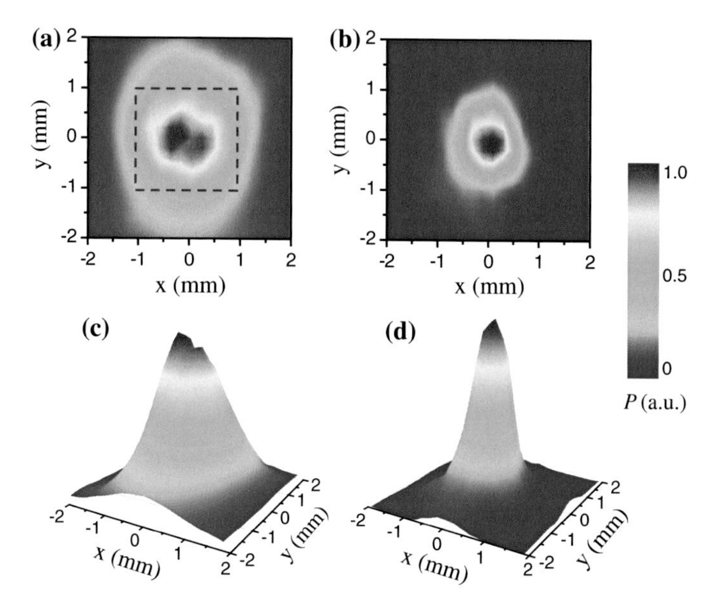

Fig. 6.2 Beam intensity profiles at the focal point of an OAP mirror raster scanned by **a**, **b** a commercial pyroelectric detector with effective detection area of 2 mm × 2 mm and **c**, **d** an AlGaN/GaN field-effect self-mixing detector with an effective detection area of 20 μm × 97 μm

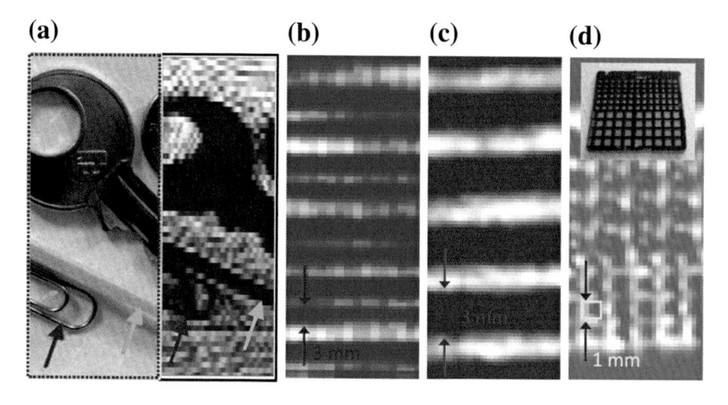

Fig. 6.3 The transmission images of different objects concealed in a paper envelope. **a** Key, match stick, and paper clip. **b** Parallel plastic strips. **c** Parallel metal strips. **d** Silicon chip with flat-bottom dents in a rectangular pyramid shape

with the optical CCD image and clearly shows the objects with different materials and structures. In the raster scan, the pixel size is fixed as 200 μm × 200 μm and a signal-to-noise ratio about 1,000 is achieved. In Fig. 6.3b, c, parallel plastic and metal strips are clearly imaged. In Fig. 6.3d, the tiny dents on a silicon chip are imaged and compared to the optical photograph. The spatial resolution is about 1 mm.

6.1.2 *Terahertz Reflection Imaging*

The reflectance of terahertz wave by different materials and surfaces differs. Hence, a reflection-type imaging can not only see the surface morphology and surface materials, but also may see the substances beneath the surface. The setup for raster-scan imaging in reflection mode is schematically shown in Fig. 6.4. A photo of the setup is shown in Fig. B.4 in Appendix B. The terahertz wave is collected by two OAPs and focused onto the object to be imaged. The terahertz wave reflected from the object is partially reflected by the beam splitter and then collected by the third OAP before it is finally focused onto the detector. Similar to the transmission mode, the different objects are hidden in an envelope and placed on the motorized state. The terahertz source and the readout method for the detector signal are the same as those used for transmission-type imaging.

As shown in Fig. 6.5, the reflection terahertz images clearly reveal the objects with different materials and structures. A terahertz reflection image of a Chinese 1-yuan coin is shown in Fig. 6.5a. The pixel size is 200 μm $\times$ 200 μm and the terahertz image consists of 150 $\times$ 150 pixels. Through the comparison of the CCD image and the terahertz image, the peony flower and the Chinese characters '1-yuan' can be clearly revealed. The spatial resolution is better than 1 mm. Figure 6.5b shows the terahertz reflection image of an empty plastic sample box. A metal washer in the box without/with a plastic lid is imaged as shown in Fig. 6.5c, d. In both cases, the outer plastic box and the washer can be clearly imaged.

To get closer to a practical terahertz imaging application, Fig. 6.6 displays terahertz reflection image of a surgical knife on a piece of genuine leather. Again, the pixel size is set to be 200 μm $\times$ 200 μm and the image consists of 70 $\times$ 150 pixels. The image shows a very high contrast between the metallic surface and the leather. Although the reflection from the leather is low, the signal still has a high signal-to-noise ratio.

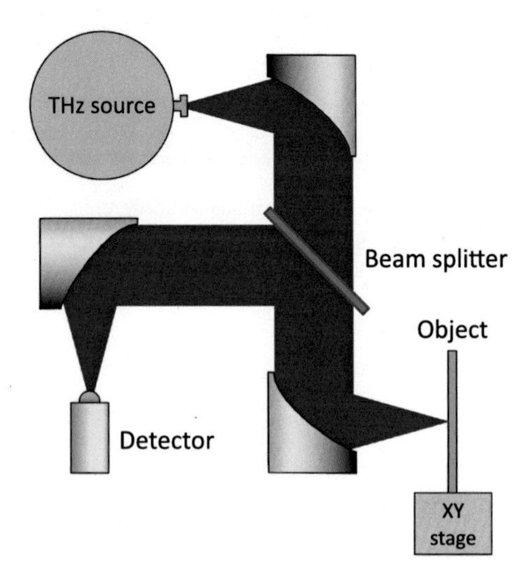

Fig. 6.4 Setup for the raster-scan imaging in reflection mode

Fig. 6.5 Reflection images of the objects with different materials and structures concealed in an envelope. **a** A Chinese 1-yuan coin. **b** Empty sample box. **c** Open plastic sample box with a metal washer on top. **d** Plastic sample box with a metal washer inside covered by a plastic lid

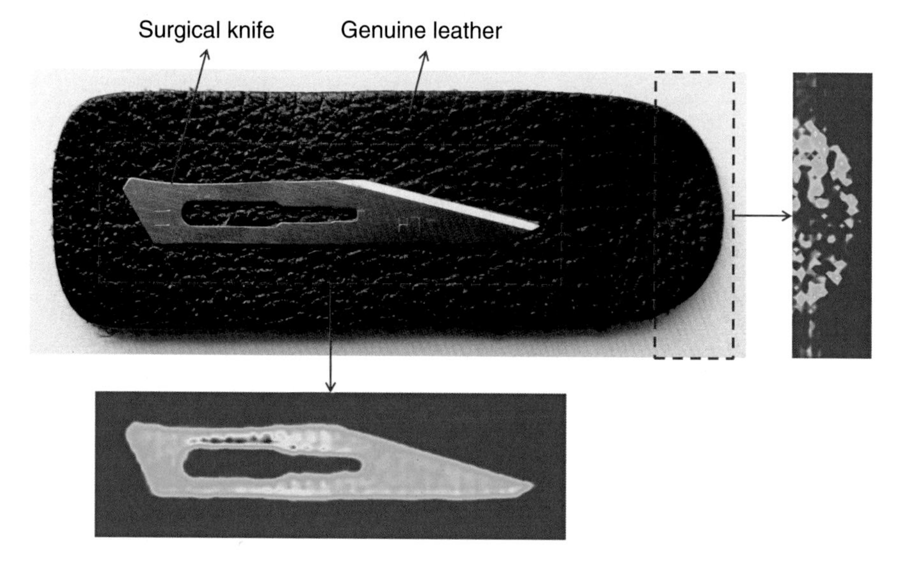

Fig. 6.6 Terahertz reflection image of a surgical knife on a piece of genuine leather

We expect that high-sensitivity self-mixing detector may play an important role in terahertz imaging applications.

6.2 Linear Detector Array

In the previous section, we have demonstrated terahertz imaging with high signal-to-noise ratio based on single-pixel self-mixing detectors. Although the response time of the detector can be made far below 1 ms, the imaging speed is limited by the slow raster scan and the slow moving linear stages. Higher speed can be achieved by increasing the speed of the linear stages and most favorably by increasing the number of detectors. Array detectors are highly desired for large-scale images. Here, we demonstrate a faster terahertz imaging using a 1×9 linear detector array, as shown in Fig. 6.7. Considering the optical diffraction limit, the pixel distance is set as 400 μm for 900 GHz terahertz imaging and the dimension of the array chip is about 4 mm $\times$ 2 mm.

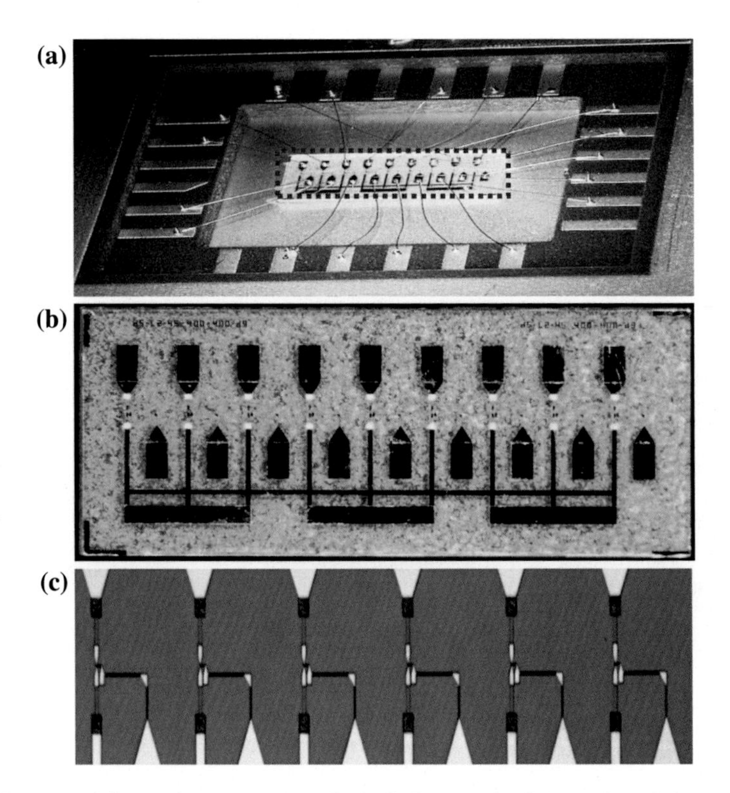

Fig. 6.7 A 1×9 linear detector array. **a** Optical photograph of the packaged detector array. **b** Linear array detector chip before packaging. **c** Zoom-in view of the pixel detectors

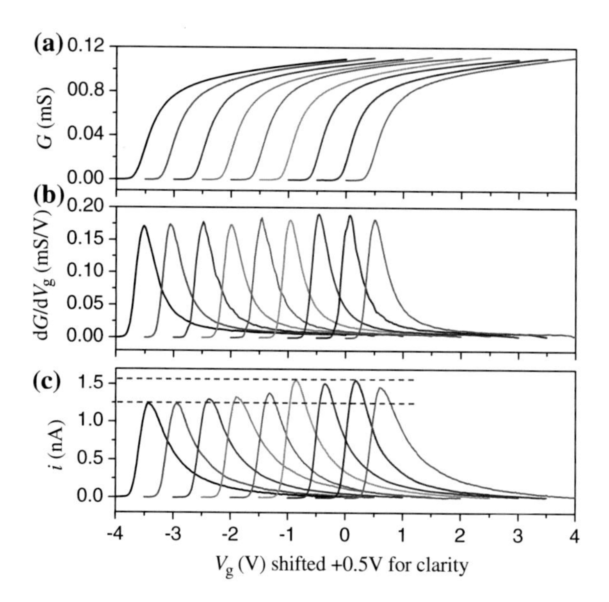

Fig. 6.8 **a** Conductance, **b** field-effect factor, and **c** the photocurrent measured as a function of the gate voltage for all of the pixel detectors. For clarity, each curve is shifted to the *right* by 0.5 V

As shown in Fig. 6.8, the source–drain conductance, the field-effect factor, and the photocurrent of the nine pixel detectors are measured as a function of the gate voltage. For clarity, the curve for each pixel detector is shifted to the right side by a gate voltage of 0.5 V. The conductance and the threshold voltage are about 0.11 mS and –3.8 V for all detectors, respectively. The electrical characteristics of the pixel detectors have a good consistency. In Fig. 6.8b, the field-effect factor for each pixel detector tuned by the gate voltage is about 0.18 mS/V with $V_g = -3.5$ V and the variation is less than $\pm 3\,\%$. As shown in Fig. 6.8c, the maximum photocurrent for each detector is around 1.4 nA at $V_g = -3.5$ V. The maximum and minimum photocurrents are marked by the dotted lines and the variation in photoresponse between the pixels is less than $\pm 10\,\%$. According to the photocurrent determined by $i_{\mathrm{T}} \propto \frac{\mathrm{d}G}{\mathrm{d}V_g} \dot{\xi}_x \dot{\xi}_z \cos\phi$, the variation in photocurrent from pixel to pixel is not only caused by the variation in the field-effect factors, but also may be induced by the unevenly distributed terahertz electric field. Although the array size is yet small in scale, it can be foreseen that large linear array and focal plane array could be developed when the quality of the 2DEG materials and uniformity of the fabrication processes are improved.

6.3 Detectors for Fourier Transform Spectroscopy

Because of the unique spectral characteristics, terahertz spectroscopy is used for the condensed-matter and medicine analysis. Femtosecond lasers enable now terahertz time-domain spectroscopy (THz-TDS) and the technology is already widely used and yet advancing at a high speed. Fourier transform spectroscopy (FTS) as the earlier

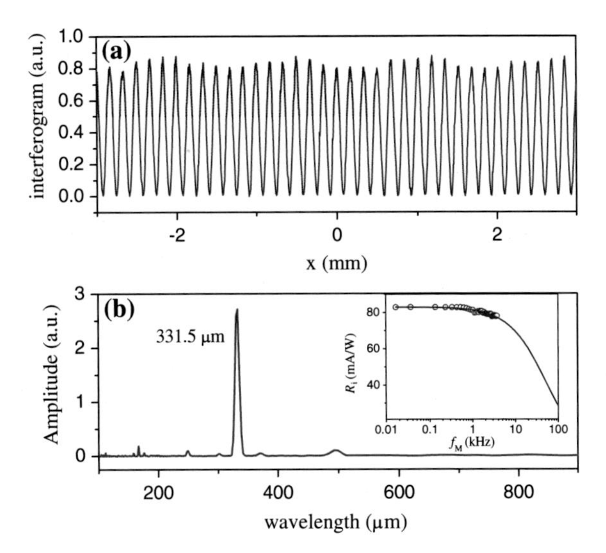

Fig. 6.9 **a** Interferogram of the terahertz signal at 905 GHz from the BWO source. **b** Fourier-transformed terahertz spectrum. The inset is the terahertz response as a function of the modulation frequency

mainstream in far-infrared spectroscopy is playing less important role than the THz-TDS due to the lack of high-power broadband terahertz source and high-sensitivity terahertz detectors for room-temperature operation.

Self-mixing detectors can be optimized to offer high sensitivity in a wide spectral range. Especially, when the coherent terahertz sources in a multiple spectral range are equipped for FTS, self-mixing detectors are most favorable for system integration. Such a new FTS will be compact, lightweight, low cost, and yet offers a high speed and a high signal-to-noise ratio. Here, we demonstrate the application of a self-mixing field-effect detector in a FTS for the analysis of a 905 GHz terahertz signal from the BWO source.

As shown in Fig. 6.9, the interferogram and the Fourier-transformed spectra are obtained. The spectrum recovers the signal at 905 GHz corresponding to a wavelength of 331.5 μm. The inset shows the terahertz response as a function of the modulation frequency indicating that the detector has a high-speed response for fast FTS measurement.

6.4 Summary

In this chapter, we have demonstrated the possible terahertz imaging applications using single-pixel self-mixing detectors for raster-scan imaging and a 1×9 linear detector array for line-scan imaging, respectively. A spatial resolution better than 1 mm limited by the terahertz beam size and a high signal-to-noise ratio are obtained. The application of the detector in a FTS to resolve a terahertz signal around 900 GHz from a BWO source is demonstrated. Further development to improve the sensitivity, the speed, the array scale, and the uniformity of the array detectors would allow us to bring many terahertz applications into reality.

References

1. Zhang, X.C., Xu, J.Z.: Introduction to THz Wave Photonics, p. 1. Springer, Heidelberg (2010)
2. Bründermann, E., Hübers, H.W., Kimmitt, M.F.: Terahertz Techniques, pp. 1–22. Springer, Heidelberg (2012)
3. Saeedkia, D.: Handbook of Terahertz Technology for Imaging, Sensing and Communications, pp. 91–120. Woodhead Publishing Ltd, Cambridge (2013)
4. Tonnuchi, M.: Cutting-edge terahertz technology. Nat. Photon. **1**, 97–105 (2007)
5. Jansen, C., Wietzke, S., Peters, O., Scheller, M., Vieweg, N., Salhi, M., Krumbholz, N., Jördens, C., Hochrein, T., Koch, M.: Terahertz imaging: applications and perspectives. Appl. Opt. **49**(19), E48–57 (2010)
6. Federici, J.F., Schulkin, B., Huang, F., Gary, D., Barat, R., Oliveira, F., Zimdars, D.: THz imaging and sensing for security applications explosives, weapons and drugs. Semicond. Sci. Technol. **20**(7), S266–S280 (2005)
7. Siebert, K.J., Loföer, T., Quast, H., Thomson, M., Bauer, T., Leonhardt, R., Czasch, S., Roskos, H.G.: All-optoelectronic continuous wave THz imaging for biomedical applications. Phys. Med. Biol. **47**, 3743–3748 (2002)
8. Arnone, D.D., Ciesla, C.M., Corchia, A., Egusa, S., Pepper, M., Chamberlain, J.M., Bezant, C., Linfield, E.H.: Application of terahertz (THz) technology to medical imaging. Proc. SPIE **3828**, 209 (1999)
9. Jepsen, P.U., Cooke, D.G., Koch, M.: Terahertz spectroscopy and imaging-modern techniques and applications. Laser Photon. Rev. **5**(1), 124–166 (2011)
10. Shen, Y.C., Lo, T., Taday, P.F., Cole, B.E., Tribe, W.R., Kemp, M.C.: Detection and identification of explosives using terahertz pulsed spectroscopic imaging. Appl. Phys. Lett. **86**, 24116 (2005)
11. Liu, H.B., Chen, Y.Q., Bastiaans, G.J., Zhang, X.C.: Detection and identification of explosive RDX by THz diffuse reflection spectroscopy. Opt. Express **14**(1), 415–423 (2006)
12. Fergusona, B., Wanga, S., Zhong, H., Abbottc, D., Zhanga, X.C.: Powder retection with T-ray imaging. Proc. SPIE **5070**, 7 (2003)
13. Cooper, K.B., Dengler, R.J., Llombart, N., Thomas, B., Chattopadhyay, G., Siegel, P.H.: THz imaging radar for standoff personnel screening. IEEE Trans. Terahertz Sci. Technol. **1**(1), 128–169 (2011)
14. Dobroiu, A., Yamashita, M., Ohshima, Y.N., Morita, Y., Otani, C., Kawase, K.: Terahertz imaging system based on a backward-wave oscillator. Appl. Opt. **43**(30), 5637–5646 (2004)
15. Lee, A.W.M., Qin, Q., Kumar, S., Williams, B.S., Hu, Q.: Real-time terahertz imaging over a standoff distance (25 m). Appl. Phys. Lett. **89**, 141125 (2006)
16. Rothbart, N., Richter, H., Wienold, M., Schrottke, L., Grahn, H.T.: Hübers, H.W.: Fast 2-D and 3-D terahertz imaging with a quantum-cascade laser and a scanning mirror. IEEE Trans. Terahertz Sci. Technol. **3**(5), 617–624 (2013)
17. Johnson, J.L., Dorney, T.D., Mittlemana, D.M.: Enhanced depth resolution in terahertz imaging using phase-shift interferometry. Appl. Phys. Lett. **78**(6), 835–837 (2001)
18. Chan, W.L., Charan, K., Takhar, D., Kelly, K.F., Baraniuk, R.G., Mittleman, D.M.: A single-pixel terahertz imaging system based on compressed sensing. Appl. Phys. Lett. **93**, 121105 (2008)
19. Lee, A.W., Hu, Q.: Real-time, continuous-wave terahertz imaging by use of a microbolometer focal-plane array. Opt. Lett. **30**(9), 2563–2565 (2005)

20. Kleine-Ostmann, T., Knobloch, P., Koch, M., Hoffmann, S., Breede, M., Hofmann, M., Hein, G., Pierz, K., Sperling, M., Donhuijsen, K.: Continuous-wave THz imaging. Electron. Lett. **37**(24), 1461–1463 (2001)
21. Hattori, T., Ohta, K., Rungsawang, R., Tukamoto, K.: Phase-sensitive high-speed THz imaging. J. Phys. D: Appl. Phys. **37**, 770–773 (2004)
22. Planken, P.C.M., Bakker, H.J.: Towards time-resolved THz imaging. Appl. Phys. A **78**, 465–469 (2004)
23. Minkevičius, L., Tamošiunas, V., Kašalynas, I., Seliuta, D., Valušis, G., Lisauskas, A., Boppel, S., Roskos, H.G., Köhler, K.: Terahertz heterodyne imaging with InGaAs-based bow-tie diodes. Appl. Phys. Lett. **99**,131101 (2011)
24. Sun, J.D., Sun, Y.F., Wu, D.M., Cai, Y., Qin, H., Zhang, B.S.: High-responsivity, low-noise, room-temperature, self-mixing terahertz detector realized using floating antennas on a GaN-based field-effect transistor. Appl. Phys. Lett. **100**, 013506 (2012)
25. Vicarelli, L., Vitiello, M.S., Coquillat, D., Lombardo, A., Ferrari, A.C., Knap, W., Polini, M., Pellegrini, V., Tredicucci, A.: Graphene field-effect transistors as room-temperature terahertz detectors. Nat. Mater. **11**, 865–871 (2012)
26. Schuster, F., Coquillat, D., Videlier, H., Sakowicz, M., Teppe, F., Dussopt, L., Giffard, B., Skotnicki, T., Knap, W.: Broadband terahertz imaging with highly sensitive silicon CMOS detectors. Opt. Express **19**(8), 7827–7832 (2011)
27. Hu, S., Xiong, Y.Z., Zhang, B., Wang, L., Lim, T.G., Je, M., Madihian, M.: A SiGe BiCMOS transmitter/receiver chipset with on-chip SIW antennas for terahertz applications. IEEE J. Solid-State Circuits **47**, 2654 (2012)
28. Knap, W., Dyakonov, M., Coquillat, D., Teppe, F., Dyakonova, N., Lsakowski, J., Karpierz, K., Sakowicz, M., Valusis, G., Seliuta, D., Kasalynas, I., Fatimy, A.E., Meziani, Y.M., Otsuji, T.: Field effect transistors for terahertz detection: physics and first imaging applications. J. Infrared Milli Terahz Waves **30**, 1319–1337 (2009)
29. Kachorovskii, V.Y., Shur, M.S.: Field effect transistor as ultrafast detector of modulated terahertz radiation. Solid State Electron. **52**(2), 182–185 (2008)
30. Boppel, S., Lisauskas, A., Mundt, M., Seliuta, D., Minkevičius, L., Kašalynas, I., Valušis, G., Mittendorff, M., Winnerl, S., Krozer, V., Roskos, H.G.: CMOS integrated antenna-coupled field-effect transistors for the detection of radiation from 0.2 to 4.3 THz. IEEE Trans. Microwave Theory Techn. **60**, 3834 (2012)
31. Bauer, M., Venckevičius, R., Kašalynas, I., Boppel, S., Mundt, M., Minkevičius, L., Lisauskas, A., Valušis, G., Krozer, V., Roskos, H.G.: Antenna-coupled field-effect transistors for multi-spectral terahertz imaging up to 4.25 THz. Opt. Express **22**(16), 19250–19256 (2014)
32. Lisauskas, A., Bauer, M., Boppel, S., Mundt, M., Khamaisi, B., Socher, E., Venckevičius, R., Minkevičius, L., Kašalynas, I., Seliuta, D., Valuis, G., Krozer, V., Roskos, H.G.: Exploration of terahertz imaging with silicon MOSFETs. J. Infrared Milli Terahz Waves **35**(1), 63–80 (2014)
33. Teppea, F., Veksler, D., Kachorovski, V.Y., Dmitriev, A.P., Xie, X., Zhang, X.C., Rumyantsev, S., Knap, W., Shur, M.S.: Plasma wave resonant detection of femtosecond pulsed terahertz radiation by a nanometer field-effect transistor. Appl. Phys. Lett. **87**, 022102 (2005)
34. Preu, S., Kim, S., Verma, R., Burke, P.G., Vinh, N.Q., Sherwin, M.S., Gossard, A.C.: Terahertz detection by a homodyne field effect transistor multiplicative mixer. IEEE Trans. Terahertz Sci. Technol. **2**(3), 278–283 (2012)
35. Öefors, E., Lisauskas, A., Glaab, D., Roskos, H.G., Pfeiffer, U.R.: Minkevičius, L., Kašalynas, I., Seliuta, D., Valušis, G., Krozer, V., Roskos, H.G.: Terahertz imaging detectors in CMOS technology. J. Infrared Milli Terahz Waves **30**, 1269–1280 (2014)
36. Sun, J.D., Qin, H., Lewis, R.A., Sun, Y.F., Zhang, X.Y., Cai, Y., Wu, D.M., Zhang, B.S.: Probing and modelling the localized self-mixing in a GaN/AlGaN field-effect terahertz detector. Appl. Phys. Lett. **100**, 173513 (2012)

37. Hadi, R.A., Sherry, H., Grzyb, J., Zhao, Y., Förster, W., Keller, H.M., Cathelinand, A., Kaiser, A., Pfeiffer, U.R.: A 1 k-pixel video camera for 0.7-1.1 terahertz imaging applications in 65 nm CMOS. IEEE J. Solid-State Circuits **47**, 2999 (2012)
38. Zdanevičius, J., Bauer, M., Boppel, S., Palenskis, V., Lisauskas, A., Krozer, V., Roskos, H.G.: Camera for high-speed THz imaging. J. Infrared Milli Terahz Waves (2015). doi:10.1007/s10762-015-0169-1

Chapter 7
Conclusions and Outlook

Abstract In this chapter, we emphasize that plasmon devices based on 2DEG materials ways become more and more important for terahertz applications. The main results and the conclusions are listed for non-resonant self-mixing detection of terahertz electromagnetic wave. The future trends are developing heterodyne plasmon detectors and resonant plasmon detectors are discussed. Questions concerning the ultimate limit in plasmon detection are raised for further studies.

The light-matter interactions in the terahertz frequency regime carry abundant material information which could lead to various applications. The key technologies of terahertz sources and detectors are the main research focuses in this field. Direct approaches toward terahertz electronics and terahertz photonics require electronic materials with superiorly high electron mobility or precisely engineered quantum level systems, ultra-short electron's transit time in nanoscale devices, and elimination of parasitic circuit parameters and/or cryogenic temperature. As a collective excitation in dense electron gas, *plasma waves* could render a new approach for the realization of room-temperature terahertz devices.

This thesis aims to develop high-sensitivity terahertz detectors based on plasma-wave excitation in gate-controlled two-dimensional electron systems at room temperature. Terahertz antennas and Schottky gates integrated with the two-dimensional electron gas (2DEG) of an AlGaN/GaN heterostructure are utilized to manipulate the localized terahertz field and the field-effect electron channel, respectively. The underlying physics of terahertz detection is uncovered through the transport studies by manipulating the localized terahertz field, the field-effect electron channel, and the coupling in between.

The following results and conclusions are obtained:

1. **Quasi-static self-mixing detector model** [1–5] The model takes into account the spatial distributions of the terahertz field and the electron density in the antenna-coupled and antenna-gated electron channel. The terahertz photocurrent is proportional to the field-effect factor $\mathrm{d}G/\mathrm{d}V_g$ or $\mathrm{d}n/\mathrm{d}V_g$ and the self-mixing factor

J. Sun, *Field-effect Self-mixing Terahertz Detectors*, Springer Theses,
DOI 10.1007/978-3-662-48681-8_7

$\dot{\xi}_x\dot{\xi}_z\cos\phi$. The model offers a complete description of the terahertz response in the two-dimensional space of the gate voltage and the source–drain bias covering the linear regime, the saturation regime, and the transition regime.

2. **High-efficiency asymmetric terahertz antenna** [1] An asymmetric terahertz antenna made of three dipole blocks is designed, simulated, and characterized. The antenna provides a large self-mixing factor $\dot{\xi}_x\dot{\xi}_z\cos\phi$ and hence the terahertz responsivity by enhancing and localizing the terahertz near field within a region of <200 nm near one of the edges of the gated electron channel.
3. **Verification of the quasi-static detector model and the asymmetric antenna design** [2] We directly probe the local self-mixing photocurrent by selectively depleting the 2DEG near the edges of the gated channel. Not only the magnitude, but also the polarity of the photocurrent is probed and tuned. The simulated photocurrent as a function of the source–drain bias and the gate voltage agrees well with the measured photocurrent. Non-uniform distribution of the terahertz field is confirmed. The detector model provides valuable guidance for the optimization of high sensitivity terahertz detectors based on antenna-coupled field-effect electron channel.
4. **Optimization of the terahertz antenna** [1] Five different antenna-coupled self-mixing detectors are designed, fabricated, and characterized. The asymmetric antenna made of three dipole blocks is found to be the most effective design comparing to the other four different designs. Except the dipole block integrated with the gate is connected to the bonding pad, the other two dipole blocks are floating and isolated from the source and the drain bonding pads. A meander-shaped terahertz low-pass filter is found to be able to effectively isolate the antenna from the near-by bonding pads.
5. **Demonstration of resonant plasmon detection** [6] The quasi-static self-mixing detector model is further verified by intentionally suppressing the self-mixing factor in a nanogate field-effect detector coupled with a symmetric antenna. By suppressing the strong self-mixing photocurrent, the relatively weak photocurrent induced by the resonant excitation of the cavity plasmon is observed at 77 K. The observed resonant response agrees well with the quarter-wavelength plasmon mode defined by the nanogates and can be well described by Dyakonov–Shur's shallow-water theory.
6. **Scanning near-field probe for rapid antenna characterization** [7] A scanning metallic probe and the field-effect electron channel are used to image the near-field property of the terahertz antenna which is coupled to the gated electron channel. This technique allows us to rapidly identify the different antenna blocks and find out the most effective block for self-mixing terahertz detection. Thus, the method provides an alternative way for rapid verification of terahertz antenna design. The experiment also suggests that a field-effect terahertz detector combined with a probe tip may serve as a high-sensitivity terahertz near-field sensor.
7. **High-sensitivity self-mixing terahertz detector and applications** Room-temperature high-sensitivity detectors are realized based on AlGaN/GaN 2DEG for the frequency band from 0.8 to 1.1 THz. The current responsivity and the voltage responsivity reach 71 mA/W and 3.6 kV/W, respectively. The noise equivalent

power is as low as $40\,\text{pW}/\sqrt{\text{Hz}}$. Both transmission type and reflection type terahertz imagings are demonstrated using a single-pixel detector. The uniformity in responsivity of a 1×9 linear detector array is evaluated at 900 GHz. A fast Fourier transform spectroscopy of a 905 GHz terahertz signal is demonstrated using a single-pixel detector.

Verified by our detectors based on antenna-coupled AlGaN/GaN 2DEG, the quasi-static detector model clearly describes a design rule for high responsivity and high-sensitivity self-mixing detectors. Further optimization on the antenna and the field-effect electron channel are expected to reduce the *NEP* below $1\,\text{pW}/\sqrt{\text{Hz}}$. Based on the elevated sensitivity, large-scale detector arrays are expected to play an important role in active terahertz imaging systems. Besides the sensitivity, the response speed can be improved so that self-mixing detectors as direct terahertz detectors may find high-speed terahertz applications such as terahertz communication, ranging, *etc*. Meanwhile, it has to be kept in mind that an antenna-coupled field-effect channel can be configured for heterodyne mixing of terahertz waves to allow for ultra-sensitive coherent detection.

It is found in this work that nonresonant self-mixing is a 'robust' detection mechanism which shows up whenever the field-effect factor and the mixing factor are enhanced simultaneously in the same region of the field-effect channel. On the contrary, the resonant plasmon detection occurs only when the resonance condition is met and requires a higher electron mobility and lower damping of the plasma wave. The experiment on the resonant plasmon detection suggests that it is rather a weak effect than the nonresonant self-mixing effect in current devices based on the AlGaN/GaN 2DEG and the antenna design. Further investigation is required to engineer the plasmon cavity with specific boundary conditions so that the quality factor of the plasmon cavity and hence the responsivity can be enhanced.

Above all, it is of fundamentally interesting to explore new terahertz device physics and answer questions such as 'What is the ultimate sensitivity limit for self-mixing detection?' and 'Can realistic room-temperature active plasmon device be made?' More efforts can be made to uncover the mystery of terahertz light-matter interaction. As a possible step forward, the nonresonant self-mixing mechanism and resonant plasmon detection mechanism can be applied to the newly discovered two-dimensional materials such as graphene to bring out new phenomena.

References

1. Sun, J.D., Sun, Y.F., Wu, D.M., Cai, Y., Qin, H., Zhang, B.S.: High-responsivity, low-noise, room-temperature, self-mixing terahertz detector realized using floating antennas on a GaN-based field-effect transistor. Appl. Phys. Lett. **100**, 013506 (2012)
2. Sun, J.D., Qin, H., Lewis, R.A., Sun, Y.F., Zhang, X.Y., Cai, Y., Wu, D.M., Zhang, B.S.: Probing and modelling the localized self-mixing in a GaN/AlGaN field-effect terahertz detector. Appl. Phys. Lett. **100**, 173513 (2012)

3. Sun, Y.F., Sun, J.D., Zhou, Y., Tan, R.B., Zeng, C.H., Xue, W., Qin, H., Zhang, B.S., Wu, D.M.: Room temperature GaN/AlGaN self-mixing terahertz detector enhanced by resonant antennas. Appl. Phys. Lett. **98**, 252103 (2011)
4. Sun, J.D., Sun, Y.F., Zhou, Y., Zhang, Z.P., Lin, W.K., C.H., Zeng, Wu, D.M., Zhang, B.S., Qin, H., Li, L.L., Xu, W.: Enhancement of terahertz coupling efficiency by improved antenna design in GaN/AlGaN HEMT detectors. AIP Conf. Proc. **1399**, 893 (2011)
5. Zhou, Y., Sun, J.D., Sun, Y.F., Zhang, Z.P., Lin, W.K., Lou, H.X., Zeng, C.H., Lu, M., Cai, Y., Wu, D.M., Lou, S.T., Qin, H., Zhang, B.S.: Characterization of a room temperature terahertz detector based on a GaN/AlGaN HEMT. J. Semicond. **32**(4), 064005 (2011)
6. Sun, J.D., Qin, H., Lewis, R.A., Yang, X.X., Sun, Y.F., Zhang, Z.P., Li, X.X., Zhang, X.Y., Cai, Y., Wu, D.M., Zhang, B.S.: The effect of symmetry on resonant and nonresonant photoresponses in a field-effect terahertz detector. Appl. Phys. Lett. **106**, 031119 (2015)
7. Lü, L., Sun, J.D., Lewis, R.A., Sun, Y.F., Wu, D.M., Cai, Y., Qin, H.: Mapping an on-chip terahertz antenna by a scanning near-field probe and a fixed field-effect transistor. Chin. Phys. B. **24**(2), 028504 (2015)

Appendix A
Symbols

e	Electron charge
f_0	Plasma resonant frequency
f	Incident terahertz frequency
f_{M}	Modulation frequency
f_{p}	Plasmon frequency
g_{m}	Channel transconductance
i_0	Internal self-mixing photocurrent
i_{M}	Self-mixing photocurrent
i_{R}	Resonant photocurrent
i_{T}	Total photocurrent
i_{xx}	Photocurrent induced from the horizontal field
i_{xz}	Photocurrent induced by both the horizontal and the perpendicular fields
k_{B}	Boltzmann constant
m^*	Effective electron mass
n_{s}	Electron density
q	Wave vector
τ	Momentum relaxation time
τ_{e}	Electron's relaxation time
τ_{p}	Plasmon's relaxation time
v_0	Internal self-mixing photovoltage
v_{xx}	Photovoltage induced by the horizontal field
v_{xz}	Photovoltage induced by both the horizontal and the perpendicular fields
$\bar{z}$	Effective distance between the gate and the 2DEG
A_0	Antenna factor
C_{g}	Gate capacitance per unit area
E_0	Free-space terahertz field strength
G	Measured differential channel conductance
G_0	Internal differential channel conductance
I_{ds}	DC source-drain current
I_{T}	Complex function of the incident terahertz power
L	Gate length

J. Sun, *Field-effect Self-mixing Terahertz Detectors*, Springer Theses,
DOI 10.1007/978-3-662-48681-8

L_{eff}	Effective gate length per unit area
N_i	Thermal noise current spectral density
N_{iB}	Measured noise current spectral density
N_v	Thermal noise voltage spectral density
N_{vB}	Measured noise voltage spectral density
NEP	Noise equivalent power
P_0	Incident terahertz power flux
Q_{p}	Plasmon quality factor
R_i	Current responsivity
R_v	Voltage responsivity
V_{ds}	Applied DC source-drain voltage
V_{D}	Drain bias applied at the drain side
V_{g} or V_{G}	Applied DC gate voltage
V_{geff}	Effective gate voltage
V_{sat}	Saturation voltage
V_{S}	Source bias applied at the source side
V_{th}	Threshold gate voltage
U_{a}	Potential induced by the incident terahertz wave
U_{g}	Gate voltage
U_{th}	Threshold voltage
W	Gate width
Z_0	Free-space impedance ($Z_0 = 377\ \Omega$)
Z_{s}	Characteristic input impedance in RF circuits which is usually 50 Ω
$\dot{\xi}_x$	Enhancement factor of the horizontal terahertz field
$\dot{\xi}_z$	Enhancement factor of the perpendicular terahertz field
$\xi_x E_0$	Channel potential induced by the horizontal terahertz field
$\xi_z E_0$	Channel potential induced by the perpendicular terahertz field
ϕ	Phase difference between the horizontal and the perpendicular terahertz field
ω_0	Plasma resonant angular frequency
μ	Electron mobility
ν_{s}	Electron saturation velocity
ν_{d}	Electron drift velocity
ε_{s}	Relative dielectric constant
ε_0	Dielectric constant of free-space

Appendix B
Experiment Setup

See Figs. B.1, B.2, B.3 and B.4

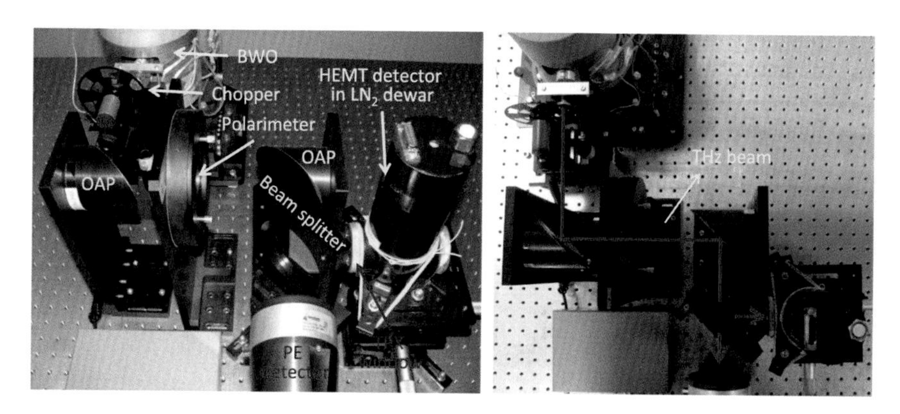

Fig. B.1 Photograph of the setup for characterizing field-effect terahertz detectors

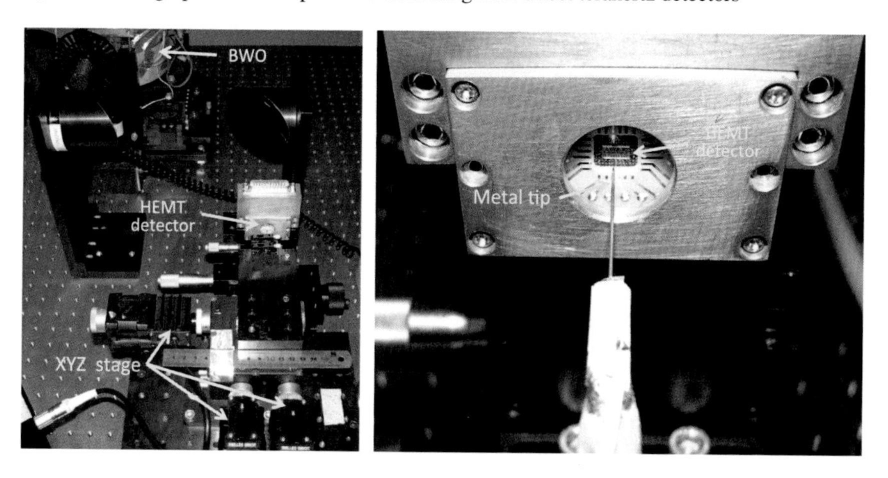

Fig. B.2 The experiment setup for scanning near-field probe of terahertz antennas

J. Sun, *Field-effect Self-mixing Terahertz Detectors*, Springer Theses,
DOI 10.1007/978-3-662-48681-8

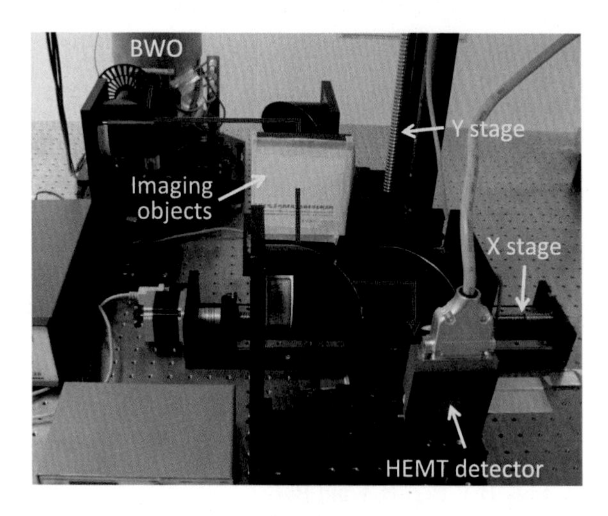

Fig. B.3 Photograph of the raster-scan imaging setup in transmission mode

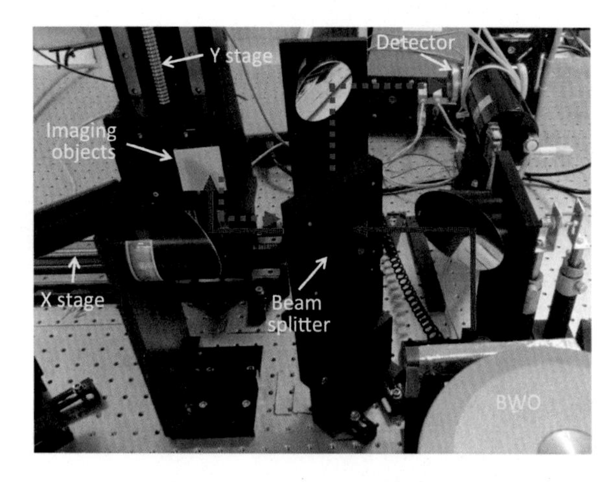

Fig. B.4 Photograph of the raster-scan imaging setup in reflection mode